DISTRIBUTED COGNITION AND REALITY

HOW PILOTS AND CREWS MAKE DECISIONS

HUMAN FACTORS AND SOCIO-TECHNICAL SYSTEMS

Series Editors:

Don Harris, Paul M. Salmon, Neville A. Stanton, and Patrick E. Waterson

Distributed Cognition and Reality: How Pilots and Crews Make Decisions
Katherine L. Plant and Neville A. Stanton

DISTRIBUTED COGNITION AND REALITY

HOW PILOTS AND CREWS MAKE DECISIONS

KATHERINE L. PLANT
NEVILLE A. STANTON

CRC Press
Taylor & Francis Group
Boca Raton London New York

CRC Press is an imprint of the
Taylor & Francis Group, an **informa** business

CRC Press
Taylor & Francis Group
6000 Broken Sound Parkway NW, Suite 300
Boca Raton, FL 33487-2742

First issued in hardback 2019

ISBN-13: 978-1-4724-8298-3 (hbk)

Library of Congress Cataloging-in-Publication Data

Names: Plant, Katherine L., author. | Stanton, Neville A. (Neville Anthony), 1960- author.
Title: Distributed cognition and reality : how pilots and crews make decisions / Katherine L. Plant and Neville A. Stanton.
Description: Boca Raton : Taylor & Francis, CRC Press, 2017. | Series: Human factors and socio-technical systems | Includes bibliographical references and index.
Identifiers: LCCN 2016014635 | ISBN 9781472482983
Subjects: LCSH: Aeronautics--Human factors. | Distributed cognition. | Airplanes--Piloting--Decision making. | Flight crews. | Group decision making.
Classification: LCC TL553.6 .P64 2017 | DDC 629.132/52019--dc23
LC record available at http://lccn.loc.gov/2016014635

Visit the Taylor & Francis Web site at
http://www.taylorandfrancis.com

and the CRC Press Web site at
http://www.crcpress.com

For 'The Plants'

Katherine L. Plant

For Maggie, Josh and Jem

Neville A. Stanton

Contents

List of Figures

List of Tables

Preface

The work presented in this book was prompted by the desire to explore the processes involved in aeronautical decision-making, rather than labelling the end product as 'pilot error'. We only have to look at recent media portrays of aviation accidents to see this. For example, in December 2014 AirAsia Flight 8501 crashed into the Java Sea after stalling in a steep climb. The pilots' responses to a mechanical failure and their apparent miscommunication were highlighted as causal factors. CNN ran a headline stating 'Pilot response led to AirAsia crash into Java Sea'. On closer examination of the details, the mechanical failure had previously occurred at least 23 times, and the pilots' initial responses were appropriate and they had followed the checklist. After repeated failure indications, the captain reset a circuit breaker which was not part of the authorised procedure. Of course, accidents arise from an interaction between technical, systemic and human factors. Operators in safety critical systems do not intentionally set out to make mistakes. Aside from the extremely rare incidents of deliberate intent to cause damage and harm, the situation surrounding the human contribution to accidents is much more complicated than it might initially appear.

The impetus for the work presented in this book is to explore aeronautical decision-making processes from the perspective of local rationality, that is why the actions and assessments made sense at the time they were made. This has been undertaken from the viewpoint of Neisser's perceptual cycle model and the associated Schema Theory. The title of this book is a direct reflection on the impact Neisser has had on the co-authors from his seminal book *Cognition and Reality*. Neville Stanton took sabbatical leave in 1998 at Cornell University in Ithaca, New York in the United States, and had the privilege to meet and dine with Professor Ulric Neisser. Ulric was very modest about his achievements. For Neville, it was a privilege to meet one of his research heroes and spend some time discussing the work on the perceptual cycle model. There is also a lineage of doctoral supervision that can be traced back to the founding father of Schema Theory, Professor Sir Frederick Bartlett. Frederick Bartlett supervised Harry Kay, Harry Kay supervised John Annett, John Annett supervised Rob Stammers, Rob Stammers supervised Neville Stanton and Neville Stanton supervised Katie Plant. We hope that we have done the work of the founding fathers justice. Incidentally, Neville was awarded the Sir Frederick Bartlett Medal by the Institute of Ergonomics and Human Factors in 2012 for his lifetime contributions to the discipline, which seems to close the research loop.

This book presents theoretical and methodological contributions. It is aimed at practitioners involved in accident and incident investigations and decision-making training, researchers and students within the disciplines of aviation, human factors, ergonomics, engineering, computer science and psychology. We hope that practitioners will make use of the methods described here and will also benefit from both the theoretical and empirical findings presented. At the same time, we expect that researchers and students will find the theoretical concepts useful and that the ideas presented here will stimulate further research in this area.

Acknowledgements

We thank the pilots who participated in the various interviews that were undertaken throughout this work. We especially thank the chief pilot at the Coastguard Helicopter Unit, who enabled access to his Search and Rescue crews. He shall remain nameless to uphold the anonymity throughout the rest of this book, but gaining this access was an integral part of the research and without it the work would not have been completed. We thank Nicholas Plant for designing the cover art.

Authors

Dr. Katherine L. Plant is a New Frontiers Fellow in Human Factors Engineering within the Transportation Research Group at the University of Southampton, Southampton, United Kingdom. Dr. Plant has primarily worked on large European Union–funded aviation projects. These have involved the application of human factors methods to design and evaluate future flight deck technologies and aviation processes. She received her BSc (Hons) in psychology from the University of Bath, and her PhD in human factors from the University of Southampton. In 2014, her PhD research was awarded the Honourable Company of Air Pilots Trust prize for aviation safety research.

Professor Neville A. Stanton, PhD, DSc, is a chartered psychologist, chartered engineer and a chartered ergonomist. He holds the chair in human factors in the Faculty of Engineering and the Environment at the University of Southampton, Southampton, United Kingdom. Prof. Stanton has degrees in psychology, applied psychology and human factors engineering and has worked at the Universities of Aston, Brunel, Cornell and MIT. His research interests include analysing, modelling and predicting human performance in systems as well as designing the interfaces between humans and technology. He has published 30 books and more than 250 journal papers on *Ergonomics and Human Factors* and is currently an editor of the peer-reviewed journal *Ergonomics*. In 1998 he was awarded the Institution of Electrical Engineers Divisional Premium Award. The Institute of Ergonomics and Human Factors awarded him the Otto Edholm Medal in 2001, the President's Medal in 2008 (with the Human Factors Integration Defence Technology Centre team) and the Sir Frederic Bartlett Medal in 2012 for his contribution to basic and applied ergonomics research. The Royal Aeronautical Society awarded him and his colleagues the Hodgson Prize and Bronze Medal in 2006 for his research on design-induced flight deck error. The University of Southampton has awarded him a Doctor of Science Degree in 2014 for his sustained contribution to the development and validation of human factors methods over the past 25 years.

List of Abbreviations

AAIB	Air Accidents Investigation Branch
ACDM	Aeronautical critical decision-making
ADM	Aeronautical decision-making
AH	Abstraction Hierarchy
ARCC	Aeronautical Rescue Coordination Centre
ATC	Air traffic control
BHA	British Helicopter Association
CAA	Civil Aviation Authority
CAL	Confined area landing
CDM	Critical decision method
CG	Coastguard
CSE	Cognitive systems engineering
CTA	Cognitive task analysis
DSA	Distributed situation awareness
EAST	Event analysis of systemic teamwork
EIS	Engine Instrument System
EOT	Engine oil temperature
FAA	Federal Aviation Authority
FDR	Flight data recorder
GC	Ground crew
GEMS	Generic error modelling system
HTA	Hierarchical task analysis
JSC	Joint cognitive system
MMSL	Manual multi-sector load
NDM	Naturalistic decision-making
PCM	Perceptual cycle model
PF(C)	Pilot flying (captain)
PNF(C)	Pilot not flying (captain)
RPD	Recognition primed decision
SA	Situation awareness
SAR	Search and rescue
SAW	Schema action world
SBATT	Suitability, barriers, approach, touchdown and takeoff
SOP	Standard operating procedure
SWARM	Schema world action research method
TDM	Team decision-making
WM	Winch man
WO	Winch operator

1 Introduction

1.1 BACKGROUND

It has been argued that advances in automation and technology have increased, rather than reduced, cognitive demands on humans (Militello and Hutton, 1998). In the highly automated aviation environment, more procedural and predictable tasks are handled by machines, whilst humans are left responsible for tasks that require diagnoses, judgement and decision-making (Militello and Hutton, 1998). This judgement and decision-making in the handling of emergency situations is usually the deciding factor as to whether an incident turns into an accident (McFadden and Towell, 1999). Decisional errors have consistently been found to account for a high proportion of pilot error (Diehl, 1991; O'Hare et al., 1994; Orasanu and Martin, 1998; Shappell and Wiegmann, 2009). Decisional factors are usually lumped together with other human factors under the umbrella term of 'human error', which is consistently implicated as a major contributor to accidents in safety critical systems and is seen by many as the principal threat to flight safety (Harris and Li, 2010).

Accidents often arise from an interaction between technical, systemic and human factors. What stands out, and is intensified by the media, is the human element as there is a desire to ascribe blame (Woods et al., 2010). However, human operators in safety critical systems do not intentionally set out to make mistakes. Aside from the extremely rare incidents of deliberate intent to cause damage and harm, the situation surrounding the human contribution to accidents is much more complicated than it might initially appear. The contemporary perspective of human error does not blame individuals or use the term as a causal attribute. Instead, human error is considered the starting point for any investigation, rejecting the notion of faulty reasoning, and seeks to explore *why* certain decisions were made over others. Dekker (2006, p. 68) summarised this:

> Human error is not an explanation of failure, it demands an explanation.

With that in mind, human error cannot begin to be understood without understanding the situation and precursors to erroneous actions, that is the decision-making processes that underpinned them. Traditional decision-making research has focused on the output product of whether a good or bad decision was made to establish the effectiveness of decision-making (Orasanu and Martin, 1998). However, understanding whether an effective decision-making process was employed, regardless of the observable manifestation of the decision, is arguably more important in order that potential training and mitigation strategies can be proposed. Furthermore, it is only with the benefit of hindsight that a label of 'bad decision-making' or 'human error' can be prescribed. What should be of interest to researchers and accident investigators is achieving an understanding of *why* actions and assessments made sense to an operator at the time they were made (i.e. local rationality).

The Federal Aviation Authority (FAA, 1991, p. ii) described aeronautical decision-making (ADM) as

> ...the mental process used by pilots to consistently determine the best course of action in response to a given set of circumstances. It is what a pilot intends to do based on the latest information he or she has...It is important to understand the factors that cause humans to make decisions and how the decision-making process not only works, but can be improved.

ADM forms the basis of the research presented in this book and is structured around three contexts. First, the primary focus will be from the perspective of aeronautical *critical* decision-making (ACDM), that is how decisions are made when dealing with critical incidents. Second, the research will focus on incidents as opposed to accidents. Fortunately, aeronautical accidents are rare and their nature means that first-hand accounts are difficult to obtain. It is acknowledged that incidents are the precursors to accidents and occur more often (Weigmann and von Thaden, 2003). Therefore, insights gained into ACDM will have the potential to prevent accidents and will have theoretical or practical relevance for the majority of aviators. Third, the majority of this research will be conducted in the context of rotary wing aviation.

Rotorcrafts, or helicopters, are to aeroplanes what motorcycles are to automobiles; there are fewer of them, but they have disproportionately higher accident rates (Stanton et al., 2015). Estimates suggest that accident rates for helicopters are 10 times higher than for fixed-wing operations (Nascimento et al., 2014). The operational environment for helicopters varies greatly with role, but helicopters generally operate outside of direct air traffic control, at low altitudes and under visual flying conditions (British Helicopter Association, 2014b). These operational advantages mean that helicopters are used in operational contexts that are not suitable for fixed-wing aircraft, including medical rescue over land, search and rescue (SAR) over water or mountains, rapid corporate passenger transfer, oil platform transfer, police search, television broadcasting, facility inspections and firefighting. However, this also means that they are vulnerable to accidents caused by degraded visibility conditions and their low-altitude operational environments (Greiser et al., 2014).

1.2 DISTRIBUTED COGNITION APPROACH

The work presented in this book is grounded in the context of distributed cognition. Stanton (2014) described distributed cognition as the process by which multiple individuals or teams work together in pursuit of a common goal, which comprises multiple interacting subgoals, in which technology often plays a central role to facilitate this. Hutchins' (1995a,b) view of distributed cognition takes the system as the unit of analysis, rather than the individuals within that system, arguing that the agents and artefacts within a system form a joint cognitive system as cognitive processes are distributed throughout it. As such, Hutchins (2000) proposed that cognitive processes may be distributed in three ways: (1) socially across team members, (2) distributed across internal and external information and

(3) distributed through time such that products of earlier events can transform the nature of later events.

Identifying cognition as a systemic endeavour led Stanton and colleagues to develop the idea of distributed situation awareness (DSA); this is founded upon the theoretical concepts of Schema Theory, the perceptual cycle model (PCM) and the distributed cognition approach (Stanton et al., 2006, 2009a; Salmon et al., 2008a). Stanton et al. (2015) described systems as a network of information elements, activated by tasks and belonging to different agents. Within this network, nodes are activated and deactivated as time passes in response to changes in the task, environment and interactions. This approach argues that awareness is distributed across human and technological agents involved in collaborative activity; it does not matter if humans or technology own this information just that the right information is activated and passes to the right agent at the right time. Salmon et al. (2015) argued that it is systems, not individuals, that lose situation awareness, and therefore, these systems should be the focus when attempting to improve performance. The PCM acknowledges the distributed nature of cognition by its emphasis on the internal schemata of the decision maker interacting with the external information in the environment and as such how cognition is distributed between people and the world. Whilst the distributed cognition perspective advocates the whole system as the unit of analysis, the research presented in this book initially focuses on individual pilots, broadening into a team-wide view. These are the units of analysis which are most relevant and accessible for the target audience of this book. However, we consistently encourage the reader to go beyond the individual, by deliberately avoiding terms such as 'pilot error', and instead encourage the consideration of the interaction between the person and their work environment.

1.3 AIMS AND OBJECTIVES

This research will explore rotary wing ACDM through Neisser's (1976) PCM and the associated Schema Theory in order to understand why actions and assessments made sense to the operator at the time they were made. The book aims to address whether the PCM can be used to study the processes of ACDM. This is structured around three key research objectives: first, the book will *contribute to the understanding of the role of the perception decision action cycle in ACDM*. This objective seeks to explore whether the PCM is a suitable framework to apply to gain a detailed understanding of critical decision-making processes. Within this, the role of Schema Theory, a central tenant of the PCM, will be reviewed in the context of Ergonomics research. Schema Theory has faced criticism, due to its mentalistic nature, that it cannot be considered a true theory. A literature review will establish if Schema Theory can be considered a valid theoretical explanation of behaviour. The construct validity of the PCM as an explanation of behaviour will also be explored. The PCM assumes that information processing occurs in a unidirectional cycle. It will be established whether this is an accurate representation of interaction between schemata, actions and world information. Second, *the development of elicitation and analysis methodologies* will produce an approach that is capable of eliciting and depicting the manifestation of perceptual cycle processes with qualitative data. The reliability of the approach will also be established. Third, the book will *investigate how teams interact in the perceptual-decision-action cycle.* The PCM is an individual model of cognition; however, previous research has used it as the theoretical underpinning to explain team processes such as DSA, although a team PCM has not been explored. As such, the final objective will explore team perceptual cycle processes.

1.4 STRUCTURE OF THE BOOK

The book has been constructed so that it can be read from start to finish by readers that are new to the subject area. Each chapter builds on information presented in the one before it. However, the chapters are also designed to be read individually for those that want to dip into specific areas. As such, some of the key theoretical constructs or central arguments are briefly refreshed at the appropriate point in each chapter. The central crux of this work is Neisser's (1976) PCM, and this is the foundation for the research presented in every chapter. Chapters 2 and 3 provide a theoretical overview of the PCM and the associated Schema Theory; in Chapter 3, this is applied in the context of the Kegworth plane crash. Chapter 4 identifies the critical decision method (CDM) as an interview technique to elicit perceptual cycle information. A coding scheme to thematically analyse the data is also presented via a critical incident case study. Chapter 5 explores the construct validity of the PCM by investigating the flow of information in the model with interviews from 20 helicopter pilots. Chapters 6 and 7 develop methodologies by presenting a more detailed PCM coding scheme for analysing qualitative data and developing an interview schedule specific to the PCM. Chapters 8 and 9 move away from individual pilots and explore the theoretical principles team contexts within the SAR domain. Finally, Chapter 10 summarises the work in light of the book's objectives and looks at possible lines of further enquiry. An overview of each chapter is provided in slightly more detail below.

Chapter 1: *Introduction* – This initial chapter introduces the area of ADM, the aims of the research and a summary of each chapter.

Chapter 2: *Schema theory: Past, present and future* – Schema Theory is intuitively appealing, although it has not always received positive press; critics of the approach argue that the concept is too ambiguous and vague and there are inherent difficulties associated with measuring schemata. Schema Theory and the associated PCM are the theoretical underpinnings of this book. This chapter addresses the criticisms of the theory by demonstrating how Schema Theory has been utilised in Ergonomics research, particularly in the key areas of situation awareness, naturalistic decision-making and error.

Chapter 3: *A case study of the Kegworth plane crash: Understanding local rationality with the perceptual cycle model* – Introduces the reader to the human error literature. Human error is a significant contributory factor to aviation incidents and accidents, but the most common criticism levelled at human error research is that inadequate causal explanations are provided. This chapter demonstrates how applying the PCM to understand 'error' can enable local rationality to be understood (i.e. why actions and assessments made sense at the time they were made). This is exemplified through an analysis of the Kegworth plane crash.

Chapter 4: *A pilot study: Using the perceptual cycle model and critical decision method to understand decision-making processes in the cockpit* – In this chapter, the CDM is utilised to collect data about an aeronautical critical incident. The data are qualitatively analysed using the principles of thematic analysis based on a coding scheme of the PCM. It is demonstrated that the approach can be used to understand the critical decision-making process and highlights how influential schemata can be at informing decision-making. The reliability of this approach is established, as well as the test–retest reliability of the CDM.

Chapter 5: *Examining the validity of Neisser's perceptual cycle model with accounts from critical decision-making in the cockpit* – The PCM assumes information processing occurs in a cyclical manner drawing on top-down and bottom-up influences to produce perceptual exploration and actions. However, the validity of the model has not been addressed. This chapter explores the construct validity of the PCM in the context of ACDM. Data from critical decision-making interviews were used to construct composite PCMs for different phases of dealing with critical incidents. A countercycle was discovered which has been attributed to skill-based behaviour, characteristic of experts. The practical applications and implications for future research are discussed.

Chapter 6: *The development of a perceptual cycle classification scheme* – In its current form, the PCM only provides a very high level of explanation. In this chapter, data from critical decision-making interviews were used to deconstruct the three high-level categories of the PCM. This resulted in the development of a 28-item taxonomy; the Schema-Action-World (SAW) classification scheme. In doing so, a more detailed description of ACDM was provided by demonstrating the relevance of different concepts in different phases of dealing with a critical incident.

Chapter 7: *The schema world action research method (SWARM) for understanding perceptual cycle processes* – The SAW taxonomy presented in the previous chapter was used to develop an interview technique that can be utilised to gain a more detailed understanding of the perceptual cycle process. A test–retest study is presented in which the reliability of the interview schedule is established. The method can be used by researchers and practitioners who want to gain a more detailed understanding of perceptual cycle processes and full procedural guidance is provided.

Chapter 8: *Team perceptual cycle processes* – As described in the previous chapters, the PCM has been successfully applied to explain individual decision-making; however, distributed decision-making in teams is the focus of much research as it is more relevant in understanding complex sociotechnical systems. This chapter explores team perceptual cycle processes in the context of SAR. It was demonstrated that the traditional perceptual cycle representation could not model the interconnectivity of teamwork effectively. As such, a network-based approach was employed to demonstrate the contributions of the different components of the PCM to the overall team process.

Chapter 9: *Exploring distributed cognition in search and rescue teams* – This chapter builds on the exploration of team perceptual cycle interactions presented in Chapter 8. Data were collected from communication recordings and observations undertaken on 12 training flights with SAR teams. The data were amalgamated into a representative, verified, case study of a SAR training sortie. The data were analysed using the shortened event analysis of systemic teamwork (EAST) method coupled with the SAW taxonomy. In doing so, the external validity of this classification scheme was assessed and relevant amendments were made. Additionally, a communication analysis supplemented the traditional EAST analysis. The findings of this chapter demonstrated how a SAR team functions in a distributed perceptual cycle.

Chapter 10: *Conclusions* – The final chapter summarises the research objectives in light of the findings. Consideration is given to theoretical, methodological and practical implications of the research.

2 Schema Theory
Past, Present and Future

2.1 INTRODUCTION

The origins of the term 'schema' are traceable to the writings of Plato, Aristotle and Kant (Marshall, 1995). The term was popularised by Bartlett in the 1930s, although Schema Theory, as it is known today, evolved after the onset of the Cognitive Revolution in the 1980s (Anderson, 1995). Schema Theory has proved contentious since the mid-1900s, when it emerged in the mainstream psychological literature. Schema Theory has seen its fair share of criticism, for example much debate surrounds the legitimacy of the term Schema *Theory*. However, the theory continues to generate research and in recent years has been influential in many areas of Ergonomics, including human–computer interaction (Chalmers, 2003), tool use (Baber, 2006), military applications (e.g. Stanton et al., 2006; Stewart et al., 2008; Salmon et al., 2009a) and a variety of transport domains including road (Hole, 2007; Walker et al., 2011; Salmon et al., 2014a), rail (Stanton and Walker, 2011; Salmon et al., 2013a) and aviation (Plant and Stanton, 2012a; Chapter 3). This chapter provides an overview of the development of Schema Theory by discussing what constitutes a schema and the pioneers of the theory. Section 2.2 discusses the criticisms that have been levelled against the theory. The chapter is intended to demonstrate how Schema Theory has influenced research questions and been applied in Ergonomics as a theoretical foundation, most notably in the fields of situation awareness (SA), decision-making and error research. In doing so, Section 2.3 demonstrates how the criticisms of Schema Theory can and have been addressed so that the value of Schema Theory can be realised in Ergonomics research and practice. Furthermore, potential future directions for Schema Theory are discussed in Section 2.4. The chapter is concluded in Section 2.5, where it is argued that the value of Schema Theory lies in its inclusion in the perceptual cycle model (PCM), which models the interaction between people and their environment, thereby providing a systemic explanation for decision-making and action. This will form the theoretical underpinnings for the remainder of the research in this book.

2.1.1 What Is a Schema?

Before discussing the role of Schema Theory in Ergonomics, it is important to clarify what is being referred to when making reference to the class of mental representation termed schemata. It is beyond the scope of this chapter to provide a review of all mental representations (see Richardson and Ball (2009) for a comprehensive review); however, a brief distinction is provided here between a schema and a mental model, the latter being the most commonly cited class of mental representation (Richardson and Ball, 2009).

Brewer (1987, cited in Stanton and Young, 2000) attempted to formalise the distinction between schemata and mental models, arguing that the former are generic mental structures which underlie knowledge and skill, whereas the latter are inferred representations of the specific state of affairs. Similarly, Johnson-Laird (1983) suggested that mental models are structural analogues of physical objects or state of affairs in the world. Richardson and Ball (2009) have argued that mental models reflect spatial information. Moray (1996) argued that the differences in the terminology for mental models did not represent differences in the concept, but instead differences in the contexts in which the use of mental models is described. It is beyond the scope of this chapter to provide further detail on mental models; interested readers are directed to Moray's (1996, 1999) classic work on mental models and Revell and Stanton's (2012) comparison of mental model concepts.

Schemata are argued to be generic knowledge structures in memory that serve to guide interpretation of external information (Grasser and Nakamura, 1982). It is hypothesised that these knowledge structures are represented as configurations of nodes linked to one another in a network of associations (Marshall, 1995). Mandler (1984, p. 14) described these as

> hierarchically organised sets of units describing generalised knowledge about an event or scene sequence.

The hierarchal nature of schemata is important; one of the key assumptions about their functioning is that actions need only be specified at the highest level schemata, and once this activation has occurred, lower-level schemata complete the action sequence relatively autonomously (Norman, 1981). It is worth noting that the nodes are not intended as physical structures, rather mental nodes and associations are used as a metaphor for the potential structure and linkage of knowledge. Grasser and Nakamura (1982) argued that schemata are generic because they provide a summary of attributes and relationships that typically occur in specific exemplars. For example, a 'restaurant eating schema' will be generic to any restaurant situation, whereas a mental model for the same situation would be concerned with eating in a specific restaurant, at a specific time.

There are a variety of definitions for the term schema (see Section 2.1), but generally a schema can be considered an organised mental pattern of thoughts or behaviours to help organise world knowledge (Neisser, 1976). When a person carries out a task, schemata affect and direct how they perceive information in the world, how this information is stored and then activated to provide them with past experiences and the knowledge about the actions required for a specific task (Neisser, 1976). Neisser (1976) proposed that there were both genotype and phenotype schemata: the former being the wider systemic factors that influence the development of individual cognitive phenomena and behaviour and the latter being the local, individual-specific manifestations of the genotype schemata. From this description, it appears that mental models are akin to phenotype schemata, whereas the traditional use of the term schema is referring to genotype schemata. Genotype schemata are triggered by the task-relevant nature of task performance. It is during this task performance that phenotype schemata (mental models) are utilised, and it is these that can be inferred from performance data (Stanton et al., 2009).

Anderson (1995) has argued that the modern use of the term schema is borrowed from the computer science and artificial intelligence disciplines that refer to data structures as schema. The dawn of the Cognitive Revolution in the 1970s saw the development of testable inferences about human mental processes, and with this, contemporary schema theories emerged, which are generally credited to Minsky's (1975) Frame Theory. Minsky argued that if machines were to be designed to carry out high-level tasks akin to human processing they would have to be provided with a lot of background knowledge as much of humans' intellectual ability comes from top-down processing using stored information. This led Minsky to introduce the concept of frames to represent knowledge of the world in machines. The terms frames and schemata are generally synonymous, although Neisser argued that frames are static and are viewed as places to put information as opposed to active schemata which can be seen as plans for obtaining more information. Minsky proposed that frames are knowledge structures with slots to accept a range of variables. Each slot has a default value (i.e. the most common representation) which is used if no value is provided by information in the world. In addition to Minsky's Frame Theory, Shank and Abelson's (1977) Script Theory is also credited with modernising Schema Theory by generating empirical research and new interpretations of schemata (Saito, 2000). Script Theory is a way to represent procedural knowledge (as opposed to declarative knowledge represented in Frame Theory). Whilst it is largely similar to Frame Theory, the values that fill the slots must be ordered in a script. The classic example of a script is the typical sequence of events that occur when ordering food in a restaurant (e.g. being seated, reading the menu, ordering food).

2.1.2 Pioneers of Schema Theory

Frederic Bartlett (1932) is attributed with the first formal definition of Schema Theory, using the theory to account for his findings in memory research. In his seminal work, *Remembering* (1932), Bartlett demonstrated that interactions between existing knowledge and new information created distortions with the latter. He described key changes that occurred to information that was recalled after variable lengths of time, noting that accurate recall was the exception not the rule. Recall was found to be a process of reconstruction, which Bartlett attributed to the existence of mental schemata. Bartlett's research provided some of the first insights into the role that past experiences have in guiding cognitive processes and modifying environmental information. He viewed schemata as active organisations of these past experiences, akin to mental templates which are neither entirely new behaviour nor a repetition of old behaviour. Brewer (2000) argued that Bartlett's major intellectual achievement was to have seen a need for a form of mental representation to explain how individuals deal with complex world knowledge.

In the 1950s, Jean Piaget (1952) focused on schemata in the context of child development and learning. Within his work, particularly *The Origins of Intelligence in Children* (1952), Piaget emphasised the active, constructive nature of the child. He argued that schemata were the building blocks of knowledge and development and that they formed the mental framework that is created as children interact with their physical and social environments. Piaget (1952) proposed three kinds of mental structures, namely, behavioural schemata, symbolic schemata and operational schemata, and argued that the processes of assimilation and accommodation are used to create schemata and influence subsequent perception. These two complementary processes are the means through which awareness of the outside world is internalised. Piaget described cognitive assimilation as the process of integrating what is perceived in the world with pre-existing internal structures (schemata). Accommodation is any modification (adaption) of the schemata based on the influence of the environment.

Arguably one of the most influential schema theorists was Ulric Neisser. Stanton et al. (2009) credited Neisser, through his book, *Cognition and Reality* (1976), with the accolade of being the most commonly used and cited text on schemata. Bartlett was one of the first to propose that higher mental processes such as remembering, thinking and perceiving were active, selective and constructive processes rather than a passive reaction or mechanical reproduction of external stimuli (Saito, 2000). It is Neisser, however, who is acknowledged for defining the active nature of schemata and creating a model to account for how schemata are utilised in the process of perception. In Neisser's view, perception is an active, cyclical process which takes place over time, during which anticipatory schemata held by the individual interact with information in the environment, thus incorporating both bottom-up and top-down processing. Neisser's view of perception was encompassed in his perceptual cycle model (PCM), which explains the reciprocal, cyclical process of interaction between individuals and the world. In this model, schemata held by individuals serve to anticipate perception, influence decision-making and direct action. The environmental experience from the world results in the modification and updating of the schemata, which in turn influences further interaction with the environment. A more detailed explanation of the PCM (and an associated figure) is provided in Section 3.1.2. Bahrick (1984, p. 36) argued:

> …no one has contributed more to the definition or formation of cognitive psychology.

Today, Neisser's work is still the foundation of many theories and studies in Ergonomics (e.g. Salmon et al., 2009; Stanton and Walker, 2011; Walker et al., 2011) and underpins the theoretical perspective of this book.

In the 1980s, Donald Norman continued to apply Schema Theory, particularly in his error taxonomy work in which he described how inappropriate activation of schemata can result in the performance of unintended actions (errors). His research led him to propose various ways in which errors in schema activation could manifest as errors of action. Furthermore, Norman and Shallice (1986) proposed a model of human performance to account for schema-driven everyday activity in which schemata are viewed as templates for behaviour triggered by cues in the environment. Their cognitive model of attention and control proposed that for well-learned action sequences, two levels of control are possible: deliberate conscious control and automatic processing. A supervisory attentional system (SAS) relates features from the environmental situation to a set of schemata. As schemata become active, responses are selected and performed. Practice of the response results in a closer coupling of feature-to-schema-to-response, producing automatic behaviour. This approach has been termed contention scheduling as a significant element of the theory is the emphasis given to prioritising competing schemata (Cooper and Shallice, 2000). This conflict avoidance system acts through the activation and inhibition of supporting and conflicting schemata. The theory states that several schemata may be activated at one time, but selection of the target schema results from the strength of activation (i.e. threshold exceeded) along with motivations of the individual. An additional control structure is required when a novel or complex task is faced, and in these instances, controlled (attention) processes come into play to manipulate activation values. Baber (2006) stated that the notion of SAS provides a mechanism for how schemata might be used and that the theory accounts for differences in novice and expert behaviour.

2.1.3 Summary

The term schema has been used in different ways by different people. For Bartlett, schemata were a form of memory bank which contained all past experiences and could explain the reason for distortions in memories. Piaget emphasised the role of schemata in child development, seeing them as the building blocks for cognition. Neisser viewed schemata as not only the plan but also the executor of the plan, that is a pattern of action as well as a pattern for action. Norman and colleagues proposed a model of everyday activity, founded in Schema Theory, which accounts for why behaviour becomes automatic with time. The ways in which the term schema has been used may vary, but common amongst them all is the agreement that schemata represent bodies of knowledge, organised in a hierarchical network which are generally activated automatically when threshold levels are exceeded. The active nature of schemata has been widely acknowledged: activation occurs in a continuous manner with top-down and bottom-up influences. The pioneers of Schema Theory were advanced thinkers for their time, and it is testament to the theory that it still has relevant application today. However, the theory is not without its critics, and the following section discusses the main criticisms levelled against it.

2.2 CRITICISMS OF SCHEMA THEORY

Schema Theory has proved contentious since its emergence in the mainstream psychological literature in the mid-1900s. Bartlett is generally considered the pioneer of Schema Theory (Stanton and Stammers, 2008); however, even Bartlett's own obituary writers were critical of Schema Theory (e.g. Broadbent, 1970; Zangwill, 1970; Oldfield, 1972). One of the key reasons for this, even argued by advocates of the approach, is that the construct of schemata is constrained by our ability to determine what a schema actually is (Smith and Hancock, 1995; Endsley, 2000). This section will describe some of the key criticisms that Schema Theory has faced, and these will be addressed in Section 2.3 through the discussion of Ergonomics research.

2.2.1 Lack of Consensus with Definitions

Richardson and Ball (2009) have argued that there is a vast array of loose and overlapping terminology surrounding mental representations in general. Even advocates of the approach concede that

> Definition of schemas are very general and often difficult to distinguish between other types of mental structures.
>
> **Mandler (1984, p. 13)**

This is evident from the mental model and schema distinction previously provided and in relation to schema a variety of definitions can be found within the literature. For example, Norman (1981, p. 3) defined schemata as

> organised memory units.

Other early definitions described schemata as mental representations of general categories (Evans, 1967) with Rumelhart and Ortony (1977, p. 101) defining them as

> data structures for representing the generic concepts stored in memory. They exist for generalized concepts underlying objects, situations, events, sequences of events, actions, and sequences of actions.

Later definitions incorporated the idea of knowledge stores for more abstract concepts such as procedural knowledge of how to do things, rather than just stores for discrete categories of things:

> In brief, schemas are higher-order cognitive structures that have been hypothesized to underlie many aspects of human knowledge and skill. They serve a crucial role in providing an account of how old knowledge interacts with new knowledge in perception, language, thought, and memory.
>
> **Brewer and Nakamura (1984, p. 120)**

Additionally, Minsky (1975) talked about frames and Shank and Abelson (1977) discussed scripts. Grasser and Nakamura (1982) acknowledged that the lack of agreement has unsettled some researchers and has resulted in debate as to whether it is appropriate at all to use the schema construct to guide theory and research. A variety of terms that have been used to define what constitutes a schema have aggravated the criticisms levelled against the theory.

2.2.2 Lack of a Unified Theory

One of the biggest criticisms of Schema Theory is that some have argued that there is not a unitary theoretical entity that deserves the label Schema *Theory* (e.g. Lodge and McGraw, 1991; Miller, 1991). Aside from the variety of ideas that have been proposed in an attempt to define what a schema actually is Schema Theory has received a variety of applications. For example, Gender Schema Theory (Bem, 1981) explains how individuals become gendered in society. Additionally, there are motor procedures that have been called schemata; Schmidt (1975) developed a motor skills variant of the approach, termed Motor Schema Theory which applied Schema Theory to the learning of discrete motor skills. Young et al. (2003) used the concept of schemata in their Schema Therapy approach to treat personality disorders and depression. The idea being that maladaptive schemata are replaced by healthy ones. Whilst the premise is the same (i.e. use of networks of information that allow for some information to be more easily assimilated than others), a variety of applications leave Schema Theory open to criticism that it does not represent a coherent theory. For example, Mandler (1984) argued that the use of the term Schema Theory was potentially misleading as a clear theory that adheres to the scientific benchmarks of testability and falsifiability was yet to be developed.

2.2.3 Methods of Measurement and Representation

Bartlett's (1932) early method of schema elicitation was termed repeated reproduction in which he sought to learn about the common types of changes people made to remembered material with increased lapses of time. Studies devoted to schema elicitation developed with renewed interest in Schema Theory at the time of the Cognitive Revolution during the 1970s and 1980s. Recall was still a key component of these studies, the output generated clustering patterns that were inferred as a schematic memory structure (Lodge et al., 1991). Early schema elicitation studies focused on understanding schemata for both objects (e.g. Bransford and Johnson, 1972; Brewer and Treyens, 1981) and events (e.g. Anderson, 1977; Bower et al., 1979). It is beyond the scope of this chapter to provide additional detail concerning these early studies, but the interested reader is directed to Rumelhart and Ortony (1977) for a full review. Whilst much information was gleaned from these original recall and remembering studies, Anderson (1995) has argued that the results of such studies are not surprising. In these studies when participants omitted or miss-recalled information, their results were considered 'wrong' and the use of schemata was inferred. Anderson, however, reasoned that if participants heard scripts with omitted or out-of-sequence information, then it made sense for them to doubt the storyteller and assume they heard it wrong or it was read wrong. Many of the early studies employed what has been referred to as 'bizarre texts' which do activate pre-existing default schemata but shed little light on the use of schemata in normal or typical situations (Sadoski et al., 1991).

Walker et al. (2011, p. 879) have acknowledged that

> ...gaining insight into mental representations...is experimentally and conceptually challenging.

Schemata are mental constructs and therefore cannot be directly measured but only inferred through the empirical consequences of them. Observable manifestations (e.g. behaviour or communication) can be used to propose that certain mental processes (e.g. utilising a schema) have taken place (Fletcher et al., 2004). The subjective nature of measuring schemata is summarised by Kuklinski et al. (1991, p. 1344):

> different researchers, armed with different ideas and a different set of questionnaire items, will inevitably discover different schemas.

Lodge and McGraw (1991), therefore, argued that it is critical that researchers choose to focus on theoretically defensible indicators of schemata or their functioning in order to reduce the criticism levelled at the concept and the associated methodologies.

2.3 ANSWERING THE CRITICS: THE ROLE OF SCHEMA THEORY IN ERGONOMICS

The work of Bartlett was an early version of Cognitive Ergonomics, that is incorporating the influences of the mind, as opposed to just the body, in the study of work. However, it was further 50 years, in the 1980s, that Cognitive Ergonomics emerged as a discipline (Hollnagel, 2001). This consideration of human–machine interaction, rather than dividing the human and the system as separate entities, is a view taken in Neisser's PCM (1976), which emphasises the interaction between person and environment and the modification each can have on one another. Hollnagel and Woods (2005) have argued that cyclical models offer a better basis to study human–technology co-agency than traditional, sequential views of information processing. This section addresses the criticisms of Schema Theory and provides an overview of the role of Schema Theory in Ergonomics research and practice, particularly the domains of situation awareness, decision-making and error research. It is argued that the value of Schema Theory lies in its inclusion with the PCM as the operator and their schema are placed in the environment in which they work (Plant and Stanton, 2012a; Chapter 3).

2.3.1 Usable Definitions

The use of Schema Theory continues into the present day in much the same ways as it was conceptualised throughout the Cognitive Revolution of the 1970s and 1980s. Regardless of the way schemata are conceptualised, whether as an organisation of past experience, as a building block of cognition or as a frame or script, much modern research has used Schema Theory as the foundation for cognitive performance in various contexts, from tool use (Baber, 2006) to e-learning (Chalmers, 2003). In an attempt to solve the issue of variations in terminology and notions about what constitutes a schema, Mandler (1984) narrowed the scope of discussion, concentrating only on the structures that organise our temporal and spatial knowledge of objects, events and places. This approach will be taken here; that is, the term schema is taken to mean the spatiotemporal schemata that organise our world knowledge (as opposed to referring to motor skills). Furthermore, whilst definitions of schemata have been found to vary, the defining features of schemata are generally agreed upon the following (Rumelhart and Ortony, 1977; Grasser and Nakamura, 1982; Stanton et al., 2009):

1. Schemata represent generic concepts which vary in their level of abstraction, permitting a range of possibilities akin to flexible templates (e.g. a shopping schema may include options such as a food shopping schema or clothes shopping schema).
2. Schemata have variables (e.g. variables in a restaurant schema include character variables, object variables and action/goal variables).
3. Schemata contain default values (e.g. these are inferences people make, such as *the chef cooked the food*, even if this is not explicitly seen or stated).
4. Schemata are embedded within other schemata and form highly integrated structures, and as such, variables have consequences on other variables (e.g. a range of cutlery for different courses suggests impressive food).
5. Schemata are not static; rather, they actively adapt based on information received (e.g. the option of online shopping can modify the traditional shopping schema).

The pioneering work of Bartlett and the schema theorists that followed remains in the modern use of Schema Theory; however, the problems and environments to which the theory is applied have become increasingly sophisticated and complex. This includes military processes such as command and control (Stanton et al., 2006; Salmon et al., 2009), fratricide (Rafferty et al., 2010) and large sociotechnical systems (Stewart et al., 2008; Jenkins et al., 2011). The fact that Schema Theory is utilised in current research suggests a consensus in the literature that a usable definition of schema is evident. The criticisms concerning the number and nature of definitions of schema do not appear to have hindered the application of Schema Theory in current practice.

2.3.2 Testable Theory

Critics have argued that Schema Theory does not fulfil the requirements of a theory. However, it is argued here that Schema Theory does fit the general requirements of a theory as defined by Kerlinger and Lee (2000), namely, that a tentative explanation of reality is offered that generates testable hypotheses which can be empirically validated or refuted to establish how accurately reality is represented. As defined by Kerlinger and Lee (2000), a theory must be capable of both predicting and explaining world phenomena. The definitions of schema have been clarified in the preceding section: Schema Theory describes how the schematic organising structures in the mind serve as mental templates to direct attention for information that is being perceived in the world and that this world information can have a modifying effect on the schemata due to the active, adaptable nature of schemata, which in turn will effect further perception (Neisser, 1976). Within this explanation, clear predictions can be made. For example, it can be hypothesised that an individual's past experience, when mapped with information in the world, will result in their behaviour. One of the fundamental principles of Schema Theory is the recognition that experiential variations between people (e.g. culturally, professionally or socially) result in different schemata which will produce different actions and behaviours (Bartlett, 1932).

Magazzu et al. (2006) found that drivers who held both motorcycle and car licences were less likely to be involved with car–motorcycle accidents, than car-only licence holders. Walker et al. (2011) discussed this in terms of Schema Theory. Schema Theory would predict that cognitive incompatibility between car drivers and motorcyclists is more prominent in car-only licence holders because the drivers do not have the appropriate schemata to deal with different road users (Walker et al., 2011). From this example, it is clear how Schema Theory is able to generate predictions and hypotheses (e.g. different road users will have different schemata which will result in different actions and behaviours) and provide an explanation of phenomena (e.g. accident rates between different road users).

Ergonomics practitioners have previously faced the accusation of using research techniques that lack a theoretical underpinning (Hollnagel et al., 1999; Ponomarenko, 2004). There have been attempts in the literature to integrate Schema Theory with the practical application of error analysis, such as Baber and Stanton's (1994; Stanton and Baber, 1996, 2005) Task Analysis for Error Identification (TAFEI) method. TAFEI allows people to predict errors in device interaction by modelling the interaction between the user and the device. The method draws upon Schema Theory by assuming that when using devices people revert to a schema for the device or similar devices; this allows for analogical reasoning to occur, that is comparing similarities between new and previously understood devices in order to understand the novel device (Schank, 1982). The use of these schemata creates expectations, and success or failure of device interaction depends on how closely a match is created between the schema and the actual operation of the device or machine.

As Holland (1992, p. 68) affirmed:

> schemata are significant because they channel experience of the present, inform anticipation of the future and play an important role in the re(construction) of memories of the past.

Schema Theory can be used to generate hypotheses; test these hypotheses with modern research methods, interpret the results in terms of Schema Theory (e.g. Walker et al., 2011), and generate practical applications (e.g. Baber and Stanton, 1994). It is therefore argued that Schema Theory does indeed meet the requirements of a theory as defined by Kerlinger and Lee (2000). A theory, however, also needs experimental, not just theoretical, justification and with that comes the need for suitable methods to assess the theory.

2.3.3 Studying Behaviour and Inferring/Representing Schemata

The early studies of inferring schemata from recall and reproduction were valuable in defining some of the key features of schemata and understanding more about their processing mechanisms (e.g. Bransford and Johnson, 1972; Brewer and Treyens, 1981; Brewer and Nakamura, 1984). However, the artificial nature of these studies gave little away about schematic processing in the real world, which is essential if the theory is going to have any practical application or relevance. As such, more naturalistic methods, which allow for the richness of qualitative data to be elicited, are more suited for the study of schemata in the real world. Schemata are mental representations of information in the mind, and therefore, they cannot be directly observed; rather, their existence is inferred from studying behaviour. It is argued, however, that schemata are no more 'fuzzy' concepts than other mental inferences, such as attitudes and belief systems, the existence of which are not disputed (Johnston-Conover and Feldman, 1991; Lodge and McGraw, 1991).

Contemporary techniques have focused on the use of classic Ergonomics methods, such as verbal protocol analysis, coupled with the use of automatic network representation methods, for example concept maps, as a more objective way to infer schemata because the knowledge networks are automatically generated. Walker et al. (2011) demonstrated this process using the software tool Leximancer™ (Leximancer, 2011). Leximancer was used to build semantic networks using transcripts elicited from verbal protocols collected during a driving task with motorcyclists and car drivers. These networks were analysed in terms of their content and various structural dimensions. It was concluded that whilst car drivers and motorcyclists shared structural similarities there was a high level of dissimilarity in content. Further analyses revealed the enduring nature of some semantic concepts (regardless of road user) and thus were inferred as genotype schemata. Phenotype schemata were also revealed via concepts that changed depending on road user type. This study demonstrated how a reliable and automatic semantic network creation tool (Leximancer), coupled with verbal protocol transcripts, was able to provide insights into the different schemata held by different road users, in a more objectively verifiable way than traditional elicitation methods would allow.

The difficulties of eliciting and representing schemata have been a barrier to the application and utility of Schema Theory. In recent years, assisted by advancements in software tools, methods have become more refined and as such gained status which has allowed Schema Theory to gain a rightful place as a legitimate line of enquiry. Refining these methods is a challenge for future Schema Theory research and other avenues exist to explore. For example, Yu et al. (2002) demonstrated how the Abstraction Hierarchy (AH; Rasmussen, 1985) could be used as a quantifiable measure of individuals' cognitive coupling with the processes they are controlling. The AH provides a hierarchical organised model of a work domain, representing both top-down and bottom-up information. This is conceptually similar to the notion of schemata. Furthermore, the approach also revealed variances between participants; therefore, the use of the AH to model different schemata held by different people is a potentially novel approach for interpreting schema and has been previously advocated by Moray (1990) for mental model representation. The advancement of methods for inferring and representing schemata results in the practical applications and implications of schema-based research being discussed with increased confidence.

2.3.4 Practical Applications of Schema Theory in Ergonomics Research

Until now, Schema Theory has been discussed in general terms, drawing on a variety of research areas to address the criticisms levelled at the theory. Discussion will now turn to specific areas of Ergonomics research and practice where Schema Theory has been significantly applied. These areas (situation awareness, naturalistic decision-making and error) attract much attention in the research literature and further highlight the valuable role Schema Theory has played in Ergonomics research.

2.3.4.1 Situation Awareness

Situation awareness (SA) has received significant attention in recent years, and considerable theoretical advances have been made (Stanton et al., 2006; Salmon et al., 2009). Endsley (1995), who has provided significant contribution to the field of SA, described the concept as an operator's dynamic understanding of 'what is going on'. Salmon et al. (2008a) have argued that of the range of models in the literature, Endsley's information processing–based three-level model is the most popular. Endsley (2000) hypothesised that schemata play a major role in dealing with the limitations of working memory which is an important influencing factor in the accuracy and completeness of SA. Experience develops operators' schemata of the systems and environments within which they operate. For example, schemata of prototypical system states can be matched with environmental cues to provide situation classification and comprehension (Endsley, 2000). SA is considered by Endsley (2000) to be the current state of the system, that is the specific state of something now, whereas a schema is an overall, general representation.

Despite the popularity and widespread use of Endsley's three-level model, it is not accepted by all in the literature and remains a contentious issue (Salmon et al., 2008a, 2015; Stanton et al., 2015). Competing models have been proposed, and whilst they offer an alternative approach to modelling SA, they are still founded in Schema Theory. For example, Smith and Hancock (1995) based their model of SA on Neisser's (1976) PCM, which describes the interaction between an individual and the environment and the influential role schemata have on this interaction. Smith and Hancock (1995) used the PCM to explain the achievement and maintenance of SA and argued that SA resides through the interaction of person and world, rather than it being a discrete cognitive product that resides in the individual, as proposed in Endsley's information processing approach. Stanton et al. (2006) argued that the PCM offers a more comprehensive description of how SA is developed and maintained when compared with the three-level model. Whilst both models are underpinned by schemata, Stanton et al. (2006) have argued that Smith and Hancock's (1995) explanation of SA goes much further than Endsley's in explaining how schemata are developed and modified. Smith and Hancock (1995) put Schema Theory and the PCM at the centre of their SA explanation, whereas Endsley (2000) merely assigns schemata as one of many information components in a larger chain of processing.

Based on the view that SA resides in neither the individual nor the world, Stanton et al. (2006, 2009) proposed a theory of distributed SA (DSA) which is founded upon the theoretical concepts of Schema Theory (i.e. genotype and phenotype schemata), the PCM and the distributed cognition approach. This approach argues that awareness is distributed across human and technological agents involved in collaborative activity (Salmon et al., 2008a). The DSA approach has been applied widely to the study of team processes in relation to SA (Patrick and Morgan, 2010; Stanton et al., 2010; Sorensen and Stanton, 2011, 2013), and a full review of its contribution is provided by Stanton et al. (2015). From the literature presented, it is clear to see how influential Schema Theory has been in the development of many theories of SA. Based on the popularity of SA in the Ergonomics literature, the concept of schema is inherent in the field, even if it is not explicitly stated.

2.3.4.2 Naturalistic Decision-Making

Hollnagel described one of the most significant changes in modern research as being the shift from 'cognition in captivity' to 'cognition in the wild', that is studying behaviour and inferring mental processes in a realistic setting (Hutchins, 1995a). Incidentally, Bartlett (1932), the pioneer of Schema Theory, also recognised the importance of ecologically valid experiments. In his studies on memory and recall, Bartlett shunned the use of traditional nonsense syllables in favour of the type of material individuals would encounter in everyday life including meaningful stories, argumentative passages and graphical material. One of the most significant applications of this paradigm shift has been within the decision-making domain. The manner in which humans make decisions is applicable to all areas within Ergonomics and as such has received much attention throughout the history of Ergonomics (Klein, 2008). The Naturalistic Decision Making (NDM) movement (Klein et al., 1989) that emerged in the late 1980s exemplifies the research shift from 'captivity' to the 'wild' (Hutchins, 1995a; Hollnagel, 2001). There is general consensus that NDM is strongly schema driven (Klein, 1993; Marshall, 1995; Smith and Marshall, 1997). Elliott (2005, p. 15) has suggested:

> ...schemata form the lenses through which the decision maker views the problem.

The dominant theory within NDM is Klein et al.'s (1986) recognition-primed decision (RPD) model. This shows how decisions for alternative actions are derived from recognition of critical information and prior knowledge. When people need to make decisions, they quickly match the situation to the patterns (schemata) they possess, which results in extremely rapid and usually successful decisions. Based on the outcome of studies comparing expert and novice decision-making, Lipshitz and Shaul (1997) argued that the original RPD model was too simplistic, and rather than being based on formal logic, decision-making depends on the representation of a situation in schemata. This schema-driven RPD model is similar to Neisser's model of perceptual action as schemata are viewed as the drivers of situational information that is available to the decision maker. Unlike Neisser's PCM, however, the schema-driven RPD model does not include feedback loops, and as such, the modifying effect that the environment can have on the decision maker's schemata and thus perceptual exploration is overlooked.

Other researchers have highlighted the influential role schemata have in decision-making. For example, Jenkins et al. (2011) argued that decision-making activity is closely linked to domain experience and the repertoire of stored schemata. Piegorsch et al. (2006) used Schema Theory to understand how Ergonomists make decisions in relation to the prevention and control of work-related injuries. In the military domain, Paris et al. (2000a) applied a schema-based method to measuring team decision-making in a navy combat information centre. This research proposed a range of training strategies based on schema-based instruction to enhance team decision-making. Additionally, in the healthcare domain, Falzer (2004) showed how Image Theory (based on the principles of Schema Theory) could be utilised to identify inhibitors to implementing evidence-based practice.

2.3.4.3 Human Error

The achievement of increased safety levels is a fundamental objective of Ergonomics research. Accidents can and do occur in all aspects of the workplace; however, they are most noticeable when they occur in high-profile and high-reliability domains such as aviation, the nuclear industry or the military. Understanding the human contribution to accidents through human error research has been a priority for many Ergonomics researchers since the dawn of the discipline and with good reason; for example, in aviation, human error is considered to be the principal threat to flight safety (Civil Aviation Authority, 1998). Stanton and Walker (2011) have argued that Schema Theory provides a good explanation for how we act in the world and, more importantly, what can go wrong.

Many of the ideas from the pioneers of human error research, including Norman, Rasmussen and Reason, have their foundations in Schema Theory (Stanton and Salmon, 2009). These error perspectives generally complement each other and are often combined. For example, in a recent analysis of the Stockwell shooting, Jenkins et al. (2011) applied a method informed by many approaches that are underpinned by Schema Theory, including Neisser's (1976) PCM, Klein et al.'s (1986) RPD model and Rasmussen's (1983) levels of information processing. Norman (1981) presented one of the first schema-based classifications of errors known as action slips. These were defined as unintended action due to the inappropriate activation of schemata. The schema perspective can explain why errors occur as numerous opportunities for errors are demonstrated. Norman (1981) classified three types of action slips: (1) slips during the formation of an intention by misinterpreting the situation, (2) slips resulting from faulty activation of schemata caused by similar triggering conditions and (3) slips resulting from the faulty triggering of active schemata (selected at an inappropriate time). Norman's taxonomy has influenced contemporary schema-based research in a variety of Ergonomics areas including road (Stanton and Salmon, 2009), rail (Stanton and Walker, 2011) and aviation (see Chapter 3). A highlight of Norman's classification scheme is that not only are the types of errors defined, but reasons for these errors provided within the context of schema activation. The provision of causal explanations allows for mitigation strategies to be considered. It is acknowledged that there is rarely a single cause of failure, but rather a dynamic interplay of multiple contributors (Dekker, 2006; Woods et al., 2010). However, what makes sense for one person is likely to make sense for someone else (Dekker, 2006); hence, there is merit in understanding operators' perceptual processes, that is their schemata and the role of these in the occurrence of error.

Reason's (1990) Generic Error Modelling System (GEMS) draws heavily on Rasmussen's (1983) skill, rule and knowledge levels of behaviour. Skill-based behaviour occurs in familiar situations when a pre-existing schema can easily be matched; resulting behaviour is therefore relatively automatic. At the other end of the spectrum, knowledge-based behaviour occurs when there is no close match to an existing schema. This requires slow and effortful cognitive processing and therefore in time-critical situations individuals may revert to applying a schema of best fit. GEMS describes how switching occurs between different types of information processing and how each level can lead to errors. For example, Reason (1990) attributed errors at the skill-based level to monitoring failures. The key feature of this model is the assertion that when confronted with a problem humans are strongly biased to find a pre-packaged solution at the skill/rule-based levels of behaviour, *before* resorting to the more effortful knowledge-based levels. Reason (1988) has argued that predictable errors are rooted in a tendency to over-utilise schemata through the processes of similarity matching and frequency gambling. Similarity matching is the process by which stimulus inputs deliver a set of retrieval cues to the long-term memory (termed the knowledge base, where schemata reside). These cues automatically activate stored schemata and possess attributes that wholly or partially match the calling conditions communicated by the stimulus input (Reason, 1990). Calling conditions frequently fail to generate a correct answer through either insufficiently specified cues or the relevant stored schemata are incomplete. This results in the activation of a number of partially matched candidates. Reason (1990) used the term frequency gambling to describe the process by which selection from these candidates is biased in favour of the more frequently encountered items. From the description of these processes, it is clear to see that the potential to make errors is high, either through inappropriate matching or gambling. For example, Reason (1988) has argued that these similarity-based searches have the effect of 'protecting the plan' against contrary evidence, which has been termed confirmation bias, that is favouring information that confirms, rather than disconfirms, beliefs (Wason, 1960). This highlights that whilst schemata are active and able to adapt to information over time, they are less flexible in moment-to-moment changes (Richardson and Ball, 2009). Reason (1988) evolved GEMS into GEMS II, in which it was proposed that automatic searching of the knowledge base (i.e. utilising schemata at skill/rule-based levels of behaviour) occurs *simultaneously* with the serial, conscious mind work (which was originally thought to occur after the automatic resources had been exhausted). Errors are the result of interactions between these two modes, and similarity matching and frequency gambling remain the primary search mechanisms (Reason, 1988).

In recent times, there has been a shift away from researching error at the individual level of analysis to understanding error in an organisational context. Schema Theory (and the related PCM) can be utilised in order to understand the schemata held by different operators at different organisational levels of a system and understanding these schemata within the context of the work system in which they materialised. As Dekker (2003, p. 100) argued:

> In order to understand what occurred in the mind, one needs to look in the world in which the mind found itself.

This is exactly what studying error from the perspective of Schema Theory and the related PCM aims to do.

2.3.5 Summary

Both SA and NDM research have implications for the study of human error. Although a high-level overview has been provided for each of these areas, the review aimed to convey the positive contribution Schema Theory has made in stimulating Ergonomics research. As discussed at the beginning of this chapter, Schema Theory is capable of generating predictions and offering explanations. It is clear from the literature presented that many key Ergonomics studies are founded in the principles of Schema Theory and studies continue to explore the role of schemata in order to advance our knowledge of human performance.

2.4 FUTURE DIRECTIONS

This chapter has discussed the development of Schema Theory and the associated PCM in terms of its origins and its current application in Ergonomics. Brief consideration will now be given to the future of Schema Theory in Ergonomics. It is fitting that in 2008, Bartlett, the founder of Schema Theory, was chosen as the subject of the peer commentary issue on *The Future of Ergonomics*. Stanton and Stammers (2008) acknowledged the impact of Bartlett's (1932) *Remembering* and how the schema concept continues to be central and influential in research and practice. Some of the key predictions made by the commentators about the future challenges faced in Ergonomics included the ever-increasing role of technology, the growth in the variety of domains the discipline covers and the diversification of users and the associated need for adaptive technology (Cacciabue, 2008; Drury, 2008). The role of schemata in understanding human interaction in the world is relevant to these future challenges. For example, schemata will be influenced by, and will influence, interaction with technologies; the use of Schema Theory has the potential to produce domain-independent research and has implications in relation to different user groups, for example novice and experts. It is argued that that the contribution of Bartlett and the schema theorists that followed, notably Neisser (1976) and his PCM, will continue to drive Ergonomics research in three key ways.

2.4.1 As a Unifying Theory in Ergonomics

Stanton (2002) has argued that although methods and models are abundant in Ergonomics, many of these fall short of being founded in academic theory. Similarly, Hancock and Diaz (2002) have argued that the development of Ergonomics theory is essential if Ergonomics is to develop as a discipline in its own right. This chapter has demonstrated that Schema Theory has been successfully applied in core Ergonomics areas, including SA, NDM and error research. Therefore, it is suggested that Schema Theory has the potential to be the foundation as a unifying theory of Ergonomics. Hancock and Diaz (2002) identified the cover-all requirement, that is the necessity for a theory to be able to predict a variety of systems interactions in a wide range of contexts, as a major stumbling block for a unified Ergonomics theory. However, Schema Theory, coupled with the associated PCM, has the potential to provide these domain-independent insights in systems interactions (Plant and Stanton, 2012a,b; Chapter 3). It is the interplay between psychological and situational factors that governs human behaviour (Reason, 1990). Furthermore, Hancock and Diaz (2002) have argued that the duality of mind and environment is the primary source of conflict between the two dominant approaches used in the study of behaviour: the ecological perspective and the information processing perspective. The PCM, with its inclusion of Schema Theory, describes a reciprocal relationship between mind and environment, and as such, this approach has merit in bridging the current gap that exists between the ecological and information processing perspectives by considering both the psychological and situational factors that govern behaviour. As Salmon et al. (2009, p. 189) affirmed:

> ...The perception-action cycle together with Schema Theory offers a theory of everything...

Schema Theory conceivably has the potential to be the unifying theory of Ergonomics, but to do this will involve the formalisation of elicitation and representation methods, as discussed in Section 2.3.3 this is already underway.

2.4.2 As a Foundation for Studying Distributed Cognition

The distributed cognition movement identifies cognition as a systemic endeavour rather than an individual phenomenon (Hutchins, 1995a). Cognition transcends the boundaries of individual actors and becomes a function that is achieved by coordination between human and technological agents. According to Stanton et al. (2010), the PCM indicates how cognition is distributed between people and the world. In this vein, Stanton and colleagues (e.g. Stanton et al., 2006, 2010; Salmon et al., 2009) proposed the theory of DSA primarily to account for SA in teams and as a reaction against Endsley's (1995) notion of shared SA. Stanton et al. (2009) have argued that schema-based accounts of SA in collaborative environments affect existing models of SA because the use of Schema Theory as a basis results in individual team members experiencing a situation in different ways, leading to compatible awareness. This contradicts traditional accounts which suggested that teams possessed shared, thus identical, SA. Furthermore, it is argued there is significant utility in progressing from linear processing models (e.g. Endsley's three-level model) to cyclical, parallel models underpinned by Schema Theory, for example the PCM (Stanton et al., 2009). Modern sociotechnical systems are regarded as distributed cognitive systems (DCS), that is studying the whole system of people and technology acting together, rather than the human–machine distinction and the interaction between the two. Within a DCS, various levels of analysis will exist; in aviation, this is exemplified as the following levels: individual pilot, flight deck, air traffic services, airline, national aviation authorities and external environments (Hollnagel and Woods, 2005). At each level of analysis, a different DCS is found which can be seen as a system within its own right that has certain functional characteristics (Hollnagel, 2005). Schema

Theory and the PCM have traditionally been concerned with the individual level of analysis; however, this has lost relevance in the study of modern, complex sociotechnical systems. The PCM is, or conceivably can be, a systems theory. By linking the mind to action to world back to mind, the PCM allows analysis and interpretation to transcend the individual levels of a system and is therefore capable of describing interactions at all levels of a DCS and between these levels.

2.4.3 As an Avenue for Exploring the Cognition of Machines

Hollnagel and Woods (1983) launched Cognitive Systems Engineering (CSE) as an approach to understand human–machine systems from a cognitive perspective as opposed to the physical or physiological perspectives that had previously prevailed. This came at a time when machines had evolved from being an extension of human's physical capabilities to mirroring mental capabilities of humans. This evolution required machines to possess both the physical and mental functionality of humans. As modelled in the PCM, humans are not linear information processors but instead behaviour is a product of the amalgamation of data-driven (world) and knowledge-driven (schemata) information. The CSE movement proposed that a comprehensive description of machines requires the inclusion of both bottom-up and top-down opportunities for knowledge. Hollnagel and Woods (1983, p. 589) argued that it is reasonable to assume that explicitly designed machines possess internal models of their environments and call these the machine 'image of the operator'. They argued that the machine image is able to assist the machine with planning, decision-making, message formulation and message interpretation. From this description, it is clear that this 'machine image' is analogous to the functions of human schemata.

Computing metaphors were once used to understand the mind; we are now in a position where mind metaphors are utilised to understand 'intelligent' machines. In this vein, Hew (2011) discussed machine schemata in an attempt at unifying the two dominant theories of SA: Endsley's (1995) shared SA and Stanton et al.'s (2006) DSA. Hew (2011) proposed a model of SA that was applicable to both humans and machines whilst retaining the richness of human cognition versus machine computation. Much like the early CSE work, the premise was that intelligent agents in a system are just like humans in the system and therefore they can conceivably possess schemata. Hew (2011) modelled schemata in order for them to be applicable to both humans and machines. The functions of a schema are twofold: first, they drive where and how information is gathered in the environment, and second, they adapt as a result of the gathered information. It is argued that machines have the ability to demonstrate behaviour that fulfils the two functions of schemata (Hollnagel and Woods, 1983; Hew, 2011). For example, an agent that employs a set of sensors and effectors can match a set of prototypes to its tracks which trigger scripts for action. This information can then be used to modify the agent's prototypes and scripts. Hew (2011) concluded that the incorporation of Schema Theory resulted in a richer and more precise description of information transactions than the stand-alone SA theories had previously achieved. Providing machines with explicit and appropriate schemata will continue their evolution into intelligent agents.

2.5 CONCLUSIONS

Schema Theory was pioneered in an era that placed value on observable manifestations of behaviour, the theory was far ahead of its time, which resulted in both recognition and cynicism. Bartlett deserves credit for pioneering the theory, although the work that followed notably from Piaget, Neisser and Norman offered unique and distinct contributions to the development of Schema Theory. Despite significant criticism, notably surrounding what constitutes a schema, whether Schema Theory is itself a testable entity and issues surrounding the measurement of schemata, Ergonomists have utilised the theory to stimulate research. Schema Theory underpins the PCM which has formed the foundation of contemporary theories of SA and has been applied to the explanations of accidents. This chapter also sought to highlight the future directions in the application of Schema Theory in Ergonomics research and practice. It is apparent that the theory, particularly when utilised in conjunction with the PCM, has the potential to offer Ergonomics research further insights into the complex interactions between humans and their environments.

Schema Theory describes how information is represented in a manner that accounts for the complexities of the human mind. The theory faces an interesting dichotomy of being intuitively appealing but also complicated to define and measure. In relation to error research, Dekker (2003) has pointed out that much of what is recorded in the literature is the result of what *can* be observed and measured, not necessarily what *needs* to or *should* be measured to gain a comprehensive understanding of error. The same can be said about the use of Schema Theory; the difficulties need to be overcome in order to realise the explanatory power of the theory. It has been argued that the significant value of Schema Theory lies in its inclusion within the PCM and the ability of this model to explain the interaction between the human and environment. To exemplify the descriptive power of Schema Theory and the PCM, Chapter 3 analyses the decision-making processes that led to the Kegworth aviation disaster.

3 Case Study of the Kegworth Plane Crash

Understanding Local Rationality with the Perceptual Cycle Model

3.1 INTRODUCTION

In Chapter 2, the theoretical underpinnings of this research, namely, Schema Theory and the perceptual cycle model (PCM; Neisser, 1976), were reviewed. This chapter applies these theories to a case study of an aviation accident in order to test their explanatory power at providing a detailed understanding of decision-making. As described in Chapter 1, the overarching aim of this book is to gain a more process-orientated understanding of decision-making as opposed to solely focusing on the end product of the decision-making process and labelling it 'human error'. The PCM provides a systemic perspective, focusing on human activity in the context in which it occurred. This chapter explores the decisions and actions of the pilots preceding the 1989 Kegworth aviation accident. It is demonstrated that the decisions undertaken, which later appeared erroneous, were, at the time, appropriate responses to the world information the pilots were presented with and the internal mental schemata that they held.

3.1.1 Decision-Making and Error in Aviation

Chapter 2 briefly introduced the issue of human error in complex sociotechnical systems, particularly the aviation domain. Incident surveys in a variety of domains consistently attribute human error as the causal factor in 70%–80% of critical events and the aviation domain is no different (Baksteen, 1995; Hollywell, 1996; Amalberti, 2001). Human error is considered to be the principal threat to flight safety (Civil Aviation Authority, 1998; Harris and Li, 2010). There will undoubtedly be human contribution to failure at some point in complex systems as a large human effort is required to maintain safety in sociotechnical systems; therefore, error is an inevitable by-product, and it is likely that human action or failure to act will be found along any path to catastrophe (Flach et al., 2008; Woods et al., 2010).

Contemporary thinking regarding human error moves the study away from blaming an individual and, rather, attempts to understand errors within the context in which they occurred. In other words, concluding that an accident was 'caused' by human error does not actually provide a causal explanation. Instead, Dekker (2006) argued that human error should be considered the starting point of any investigation in an attempt to understand how people's assessments and actions made sense to them at the time. Similarly, the naturalistic decision-making (NDM) domain rejects the notion of faulty reasoning but instead attempts to understand *why* poor decisions were made. Literature over the past decade (Bennett, 2001; Dekker, 2006; Woods et al., 2010) concurs that people do not deliberately set out to make mistakes, particularly catastrophic ones. Operators, especially in safety critical domains such as aviation, are trying to do the most professional job they can. Decisions made by an operator are embedded in a wider context (Hall and Silva, 2008). Gaining an understanding of why certain actions were chosen is essential if any progress is to be made to ensure the same mistakes are not made again (Flach et al., 2008).

There is overwhelming agreement in the literature that any research or perspective on human error needs to take the context of the system in which operators work into account (Bennett, 2001; Dekker, 2006; Woods et al., 2010). A complex system is not a single entity but is built up of interacting layers including management, maintenance, technology and operators. Layers may have competing goals meaning trade-offs will be inevitable (Reason, 1990), and accidents often emerge due to the interdependence of systems. Baksteen (1995) likened the aviation system to a pyramid. At the topmost level is the aircraft design; the descending levels of the pyramid are activities that lead to a specific flight, including air traffic services, tanking of the fuel and maintenance schedule. Baksteen acknowledged that any level things can and do go wrong. Pilots are the penultimate level and act as a 'funnel stop' to detect the deficiencies in other levels. They can, however, also add to these failures with their own mistakes which is often the reason that errors that manifest at the sharp end of an accident are blamed on the system operator but are usually symptomatic of problems deeper within the layers of a system (Reason, 1990). As noted by Bennett (2001, p. 2):

> ...crashes are the tiny tip of an iceberg composed of hundreds...of less dramatic incidents involving an aircraft or its environment.

Dekker (2006) argued that to answer the question of 'why' an error occurred, it is essential to understand the interaction between the blunt and sharp ends of a system. This view is not new; in fact, Pidgeon and O'Leary (2000) suggested that the interaction approach was defined by Turner (1978), a decade before much of the error research began to be generated. Investigating error requires an explanation for why it made sense for an operator to do what they did and inevitably this will require a certain reliance on models and theories. This chapter proposes that Schema Theory and the PCM offer a theoretical framework for understanding how the sharp and blunt ends of a system interact to lead an operator into actions that, with hindsight, seem to be erroneous.

3.1.2 Perceptual Cycle Model

Schema Theory is discussed in detail in Chapter 2, but a refresher is provided here. For the purposes of this chapter, and the remainder of the book, a schema will be considered as an organised mental pattern of thoughts or behaviours to help organise world knowledge (Neisser, 1976). Schemata provide instruction to our cognition and organise the mass of information we have to deal with (Chalmers, 2003). Our knowledge about everything can be considered as networks of information that become activated as we experience things and function according to schematic principles (see Chapter 2). When a person carries out a task, schemata affect and direct how they perceive information in the world, how this information is stored and then activated to provide them with past experiences and the knowledge about the actions required for a specific task (Mandler, 1984). Neisser's (1976) PCM incorporates Schema Theory into his explanation of information processing. This is where the role of expectation and experience influences our active exploration of the world. The initial triggering of a schema is a bottom-up process produced from situations within the environment; these initiate schemata that are based on past experiences and expectations, at which point the process becomes a top-down approach. The main cognitive structures that determine the process are anticipatory schemata (Chimir et al., 2005). This reciprocal, cyclical nature between person and environment forms the basis of Neisser's (1976) PCM. As illustrated in Figure 3.1, Neisser presented the view that human thought is closely coupled with a person's interaction in the world, both informing each other in a reciprocal, cyclical relationship. World knowledge (schemata) leads to the anticipation of certain types of information; this then directs our behaviour to seek out certain types of information and provides a way of interpreting that information. The environmental experience results in the modification and updating of cognitive schemata, and this in turn influences further interaction with the environment.

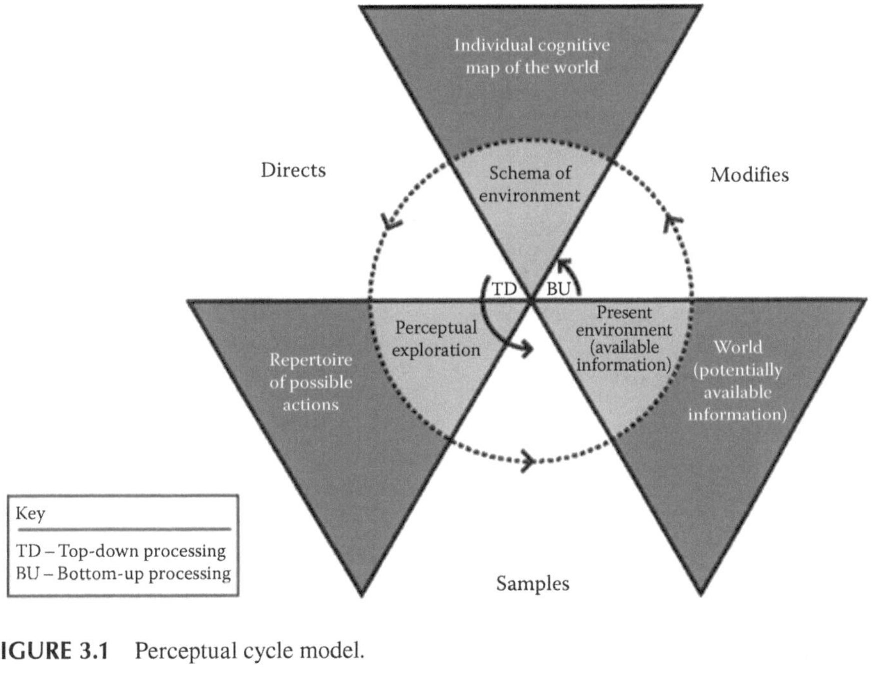

FIGURE 3.1 Perceptual cycle model.

Original NDM research throughout the 1980s concluded that people use prior experience to rapidly categorise situations, and early findings in this domain suggested people were not passively awaiting the outcome of their actions but were actively engaged in shaping events (Klein, 1998). This model places emphasis on the processes involved rather than the output products, which seems to be an appropriate stance to take when investigating decision-making and error. The product has already occurred (an erroneous action); therefore, the processes that led to it require explanation. From this perspective, decision-making can be understood by understanding the situation someone was working in, how this situation changed over time and how the evolving situation caused actions to be made that were relevant at the time. The PCM seems a relevant model to examine the *why* rather than the *what* of decision-making and error. In a cockpit environment, assuming that the pilot has the correct knowledge of the system they are operating, their schemata will enable them to anticipate events and world information will further reinforce this anticipation. This is exemplified in Figure 3.2 with an example of lowering the landing gear (read in sequence of events as denoted by numbers).

It is clear that in some circumstances, this schema may lead to inappropriate lowering of the landing gear if, for example, the plane is descending at a slower or faster

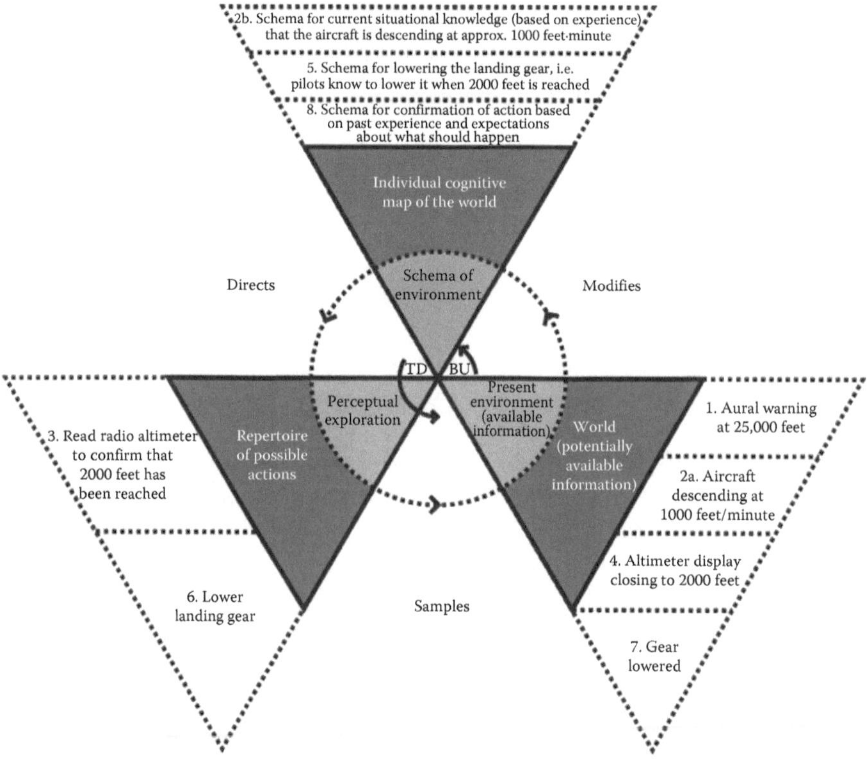

FIGURE 3.2 Perceptual cycle explanation of lowering the landing gear.

rate than the assumed 1000 feet/minute. It is up to the pilot to modify their reaction in time or else a schema-driven error will occur (a brief summary of schema-based errors is provided in the following section, but the reader is directed to Norman (1981) for a full explanation). Berman and Dismukes (2006) used the same example of lowering the landing gear to emphasise the point that experienced pilots execute actions in a largely automatic way, based on schematic representations.

The PCM offers a human-in-the-system approach to the study of cognition and action. The human-centred aspect comes from the fact that schemata are personal mental representations; it is argued that no two people will ever have precisely identical schema because they will all have different past experiences to a greater or lesser extent. Many researchers highlight the need for human-centred approaches in safety critical systems such as aviation (Shappell and Wiegmann, 2009; Stanton and Salmon, 2009; Woods et al., 2010). It is argued that although these systems are now highly automated whilst the human operator is still operating at the 'sharp end', there is a need to understand their cognitive processes that lead to decisions and actions. However, this human element must be understood in the systemic context and the world element of the PCM allows for a variety of environmental factors to be accounted for. Fedota and Parasuraman (2009) argued that it is sensible to adopt a middle stance because whilst it was correct to move away from the old view of error (i.e. operator totally to blame), there may now be too much shift towards the new view (i.e. environment totally to blame). Therefore, the interaction modelled in the perceptual cycle framework would appear a suitable foundation to model decision-making and associated errors.

3.1.3 Schema Theory and Error

Schmidt (2003) argued that one of the most interesting and useful aspects of Schema Theory is its treatment of errors as the everyday actions we produce. Norman's (1981) pioneering work presented one of the first schema-based classifications of errors known as action slips. These were defined by the performance of an unintended action due to the inappropriate activation of schemata. Norman proposed the activation-trigger-schema system to explain action, arguing that action sequences are triggered by schemata if particular conditions are satisfied. This schema perspective can explain why errors occur. The Theory of Action shows numerous opportunities for errors, such as the inappropriate activation or faulty activation of schemata. Norman classified three sources of action slips:

1. Slips during the formation of an intention by misinterpreting the situation; this produces mode errors (erroneous classification of the situation which are particularly relevant to aviation as they are the result of interaction with technology) and description errors (incomplete specification of the intention)
2. Slips resulting from faulty activation of schemata caused by similar triggering condition, producing capture error, data-driven error, association-activation error and loss-of-activation errors
3. Slips resulting from the faulty triggering of active schemata (selected at an inappropriate time)

Schema theorists would argue that schemata come into play when performing highly skilled behaviour as the behaviour occurs in the context of practised routines based on stored actions and perceptual patterns, such as piloting an aircraft (Sarter and Alexander, 2000; Stanton and Salmon, 2009). Differences in schematic representations have been repeatedly demonstrated to account for performance differences between novices and experts (Mandler, 1984). For an expert, highly skilled behaviour is almost automatic, making use of predefined schemata to guide behaviour (Berman and Dismukes, 2006). In a study of aviation accidents, Wiegmann and Boquet (2005) found skill-based errors were involved in over 80% and were the primary cause in 61%. Similarly, Lenne et al. (2008) found that skill-based errors were the most frequently cited unsafe act committed in their analysis of general aviation accidents in Australia (61%). The fact that skill-based errors are highly represented in accidents suggests that applying a theory suited to skilled behaviour is appropriate.

Ergonomics practitioners have previously faced the accusation of using research techniques that lack a theoretical underpinning (Hollnagel et al., 1999; Ponomarenko, 2004). Whilst attempts have been made to base error taxonomies on theory, such as Shappell and Wiegmann's (2000) Human Factors Analysis and Classification System method (which draws heavily on Reason's (1990) latent failure theory of error), the underlying psychological mechanisms causing errors to occur are not fully considered and the method is still a classification scheme, though arguably a sophisticated one. Dekker (2006) argued that the answers to why people do what they do lie in the context surrounding their actions (which the PCM takes account of); therefore, to achieve a causal explanation, a deeper analysis, other than classifying and counting errors, is required. In NDM research, the experienced decision maker is the object of study, experience is the basis of knowledge that the NDM domain seeks to understand and an important application of such research is to train people to achieve expertise quickly (Klein, 1998). Expertise, however, can in some cases be detrimental. Klein (1998, p. 168) defined expertise as

> Learning how to perceive.

However, judgements and decisions can be based on perceptions that are not accurate. The NDM domain defines lack of experience as a key factor in the production of errors (Lipshitz et al., 2001); therefore, studying decision-making from the perspective of the PCM will hopefully lead to insights about the role of expertise in decision-making.

In summary, schemata can be considered mental representations of the world, based on past experience and expectation that organise our knowledge of concepts and actions and can adapt to new information. From a review of the recent error literature, it would appear that there are five key points any research into human error needs to take account of:

1. Human error needs to be the starting point of research and not the concluding label (Reason, 1990; Dekker, 2002; Klein, 2008).
2. Error counting and classifying is not necessarily increasing understanding of why errors occurred nor is a causal account provided from this method (Reason, 1990; Lipshitz et al., 2001; Dekker, 2006).

3. Human error research needs to have a system perspective but still maintain a human-centred focus (Reason, 1990; Dekker, 2002; Hall and Silva, 2008; Stanton and Salmon, 2009).
4. Human error research must take the context of the work situation and the interaction between situation and mind into account (Woods et al., 2010).
5. The key question that any human error work needs to answer is *why* the actions and assessments made sense to the operators at the time (Dekker, 2006).

From the preceding discussion, it would appear that there is merit in using Schema Theory and the associated PCM as a theoretical basis for understanding decision-making and error. This will be investigated using the Kegworth aviation disaster as a case study.

3.2 KEGWORTH DISASTER

3.2.1 Synopsis

> At 18:45 on January 8th 1989 a Boeing 737-400 landed at London Heathrow Airport after completing its first shuttle from Belfast Aldergrove Airport. At 19:52 the plane left Heathrow to return to Belfast with eight crew and one hundred and eighteen passengers onboard. As the aircraft was climbing through 28,300 feet the outer panel of a blade in the fan of the No. 1 (left) engine detached. This gave rise to a series of compressor stalls in the No. 1 engine, which resulted in the airframe shuddering, smoke and fumes entering the cabin and flight deck and fluctuations of the No. 1 engine parameters. The crew made the decision to divert to East Midlands Airport. Believing that the No. 2 (right) engine had suffered damage, the crew throttled it back. The shuddering ceased as soon as the No. 2 engine was throttled back, which persuaded the crew that they had correctly dealt with the emergency, so they continued to shut it down. The No. 1 engine operated apparently normally after the initial period of vibration, however during the subsequent descent the No. 1 engine failed causing the aircraft to strike the ground 2nm from the runway. The ground impact occurred on the embankment of the M1 motorway. Forty-seven passengers died and seventy-four of the remaining seventy-nine passengers and crew suffered serious injury.*

The event was entangled with a host of human factors issues; there was confusion on the flight deck over which engine was damaged. The crew were unable to elicit the correct information from their instruments, nor did any of the passengers or cabin crew question the flight deck over their cabin address which stated the right engine was faulty even though smoke had been seen from the left. Although the Kegworth crash occurred over 20 years ago and a number of other studies have been published examining various aspects of the accident (see Booth, 1991; Johnson et al., 1995; Johnson, 1997; Besnard et al., 2004), the accident was chosen due to its prominence in British aviation history and that it allows the approach proposed here to be exemplified as it puts into context the three interacting components of the PCM (i.e. schemata, actions and world information). Furthermore, Besnard et al. (2004) argued

* The synopsis was adapted from the Official Air Accident Investigation Branch report (AAIB) (1990).

there is value in studying the psychological mechanisms that were causally involved in the Kegworth crash as psychological mechanisms are domain independent and therefore knowledge may be gained that can be applied to human–machine interaction in general. The aim of this chapter is not to reinvestigate the accident or judge previous accounts of it; rather, it is to present a PCM explanation of the decision-making that occurred during the accident.

3.2.2 Method of Analysis

The method undertaken was based upon a thematic analysis of the Kegworth plane crash. The following procedure was employed for the analysis of the Kegworth accident report (AAIB, 1990), with the intention that it can be applied to either interview data or existing accident reports.

1. *Collect data*: Either collect data through interviews (which will need to be transcribed) or obtain an accident report (in this case, the full accident report was obtained from the AAIB website).
2. *Define unit of analysis*: This chapter does not intend to make a distinction between the pilot and the co-pilot and the different schemata they may have. The flight deck is being treated as one functional unit, following a distributed cognition approach (Hutchins, 1995b). There are different levels of the system that could be analysed, for example, at the micro-level the schemata, actions and world information of the individual operators (i.e. pilot, co-pilot and cabin crew) could be analysed. Alternatively, at a more macro-level, the analysis could extend beyond the flight deck and include other aspects of the system such as ATC, passengers and company managers. It is up to the practitioner to define and set the boundary of analysis.
3. *Code data*: Thematically analyse the data using a coding scheme based on the categories of the PCM.

   Coding scheme:

   Schema: Schema of present environment (current/existing knowledge), cognitive map a person has of the world based on experience and expectations.
   Kegworth example: *...belief that the engine was on fire...*
   Action: Perceptual exploration, locomotion and action (actually doing something)
   Kegworth example: *throttled back and shut down the right engine*
   World: Actual world, potential or actual available information (physical things, conditions).
   Kegworth example: *...noise and shuddering... ceased...*
4. *Compute inter-rater reliability*: Three colleagues who were not involved in the study coded segments of the AAIB report to assess for inter-rater reliability. The analysts were supplied with a set of coding rules to guide their rating. The results of this coding were compared to that of the original analyst and a percentage of agreement between the raters was derived. Marques

and McCall (2005) provide a review of inter-rater reliability, concluding that an acceptable level of agreement can generally be defined as 80% agreement or above when reliability is calculated as the number of agreements divided by the total number of agreements and disagreements. Of the 33 segments, agreement between the three coders and the criterion coder ranged from 80% to 86% (see Chapter 4 for more detail about the analysis of reliability).

5. *Structure analysis*: To aid the explanation, it is suggested that the analysis is structured either by a timeline of events or by contributory factors (done here in line with the AAIB report). Alternatively, this could be structured around the key decisions made by the operator(s) under investigation.
6. *PCM analysis*: The transcribed and coded data can be structured around the PCM (see Figures 3.3 through 3.7). This analysis can help define recommendations and avenues of future investigation.

3.3 THEMATIC ANALYSIS OF THE KEGWORTH CRASH

As highlighted in the synopsis of the Kegworth accident, many human factors contributed to the crash. This section investigates the actions and decisions of the pilots in the Kegworth accident from the perspective of the PCM, structured around the fundamental error of shutting down the wrong engine and the five contributory factors identified in the AAIB report as the antecedents of the event.

3.3.1 Fundamental Error: Shut Down the Wrong Engine due to Inappropriate Diagnosis of Smoke Origin

> The pilots believed that smoke entering the flight deck was coming forward from the passenger cabin. Their appreciation of the air conditioning system contributed to the pilots' belief that this meant a fault with the right engine (instead of the damaged left one).

The decision made by the pilots about which engine was damaged was partly based on their assumed knowledge of the air conditioning system. This is a classic example of how people rely on their schemata. The captain's appreciation of the air conditioning system was correct for other types of aircraft flown in which he had acquired substantial flying experience. Therefore, the captain held a schema that a problem with the right engine would result in smoke in the passenger's cabin which could blow forward onto the flight deck due to the configuration of the air conditioning system. Fundamentally, this resulted in the action of the wrong engine being shut down. In a previous generation of aircraft (i.e. series 300 rather than series 400), this would be an entirely accurate schema; however, it was not appropriate for the current aircraft type. The captain reverted to his default (wrong) schema that had been developed from years of similar situations. Reason (1990) stated that errors are due to human tendency towards the familiar, similar and expected because people favour using schemata that are routine to them. Similarly, when conducting task analyses for navy tactical decision makers, Morrison et al. (1997) found that 87% of information transactions associated with situation assessment involved feature matching strategies, that is, matching the observed event to those previously

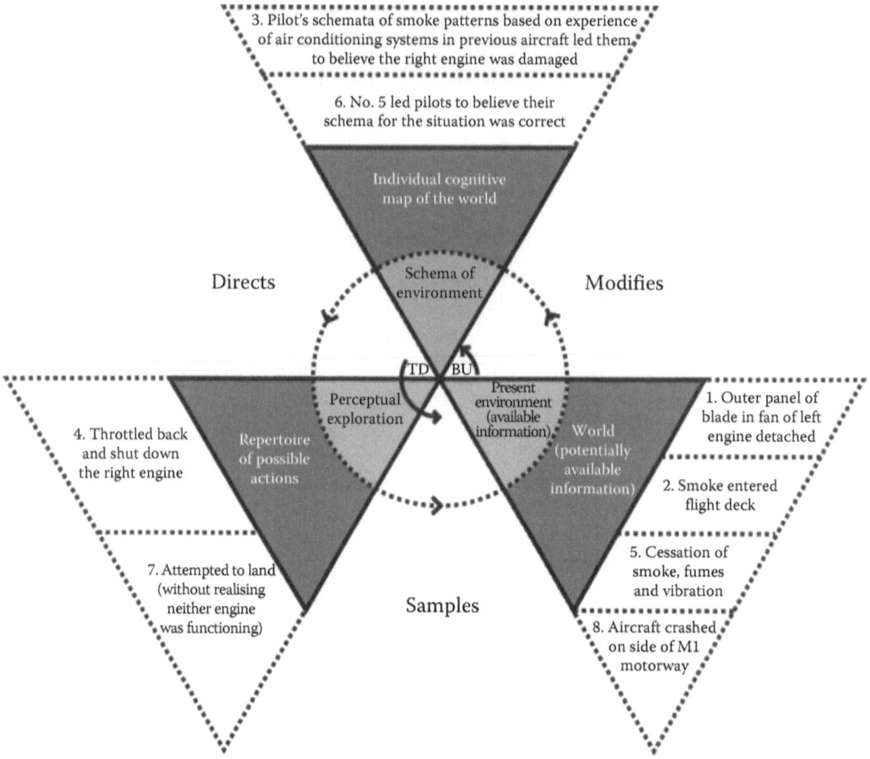

FIGURE 3.3 PCM annotated to show the fundamental error of shutting down the wrong engine.

experienced. The underlying principle of Schema Theory is the use of previous experience to develop a set of expectations, even if they subsequently turn out to be wrong. In the 400 type series, the air conditioning system did not work in the same way as the pilot's schema for the situation. In Norman's (1981) view of schema triggering components, the smoke on the flight deck thought to have come forward from the passenger cabin was enough to trigger the schema for this situation. This led to an erroneous classification of the situation. Thus, the action was intended and correct for the assumed situation (right engine damage) but not the actual situation (left engine damage). In the PCM, schemata are anticipations and are the medium by which the past affects the future; that is, information already acquired determines what will be picked up; this process is clearly evident in this example (see Figure 3.3).

3.3.2 Contributory Factor 1: The Situation Was Outside of the Crew's Training and Experience

The combination of engine vibrations, noise and smell was outside of the crew's training and experience.

The error literature describes inaccurate or incomplete schemata as 'buggy' (Groeger, 1997). It can be argued that the crew in the Kegworth accident had 'buggy' schemata for the situation they were in because they had not been in that situation before and therefore they had not built up an accurate schema of what was going on and how to deal with it. The crew, however, did have an extensive knowledge of the previous generation of the aircraft type, which in this case appeared to influence their decision-making. Kontogiannis and Malakis (2009) suggested that operators in complex sociotechnical systems need to be in a state both mindfulness and ambivalence to allow for awareness-based detection. In other words, operators are faced with partly novel (in this case, the fan blade rupture in the left engine caused smoke and vibration that was outside of the crew's training and experience) and partly familiar (in this case, the crew had an awareness from their experience of other aircraft types that smoke entering the flight deck from the cabin was likely to mean a problem with the right engine) situations. However, this dual state of belief and doubt is hard to achieve, especially in time-critical situations.

The incident of the fan blade rupture and its associated symptoms (fumes, vibration) was a rare event and not included in training, nor had the crew any firsthand experience of it, and therefore, they did not have the schema available to deal with it adequately. Dekker (2006) discussed the process of sense-making when faced with unfamiliar or unexpected problems and how action (i.e. doing something rather than doing nothing) simplifies the diagnostic problem because sense-making commits the operator to an interpretation of the event. However, this can then lead to a form of cognitive fixation where the person's perceptual exploration is biased in a certain direction which can activate certain schemata at the expense of others. As this situation was outside of the crew's training and experiences, it is likely that the processes discussed by Dekker (2006) occurred with the crew on the Kegworth flight. Similarly, Stanton and Walker (2011) found that prior experience of a particular gantry, combined with the driver's training and experience, played a role in the actions of the train driver that led to the Ladbroke Grove rail crash. Figure 3.4 sums up contributory factor 1 in the PCM.

3.3.3 Contributory Factor 2: Premature Reaction to the Problem

> The speed at which the pilots acted was contrary to their training and the Standard Operating Procedures.

Norman and Shallice (1986) further developed the ideas of Schema Theory when creating their cognitive model of attention and control, which distinguished between automatic and willed control. Within this model, they argue that schemata are templates for behaviours triggered by cues in the environment. Similarly, the Cognitive-Regulation Levels of Action Theory (Doos et al., 2004) stated that human action takes place at different cognitive levels; distinguishing between conscious attention-requiring action and unconscious action that does not require the presence of thought, this latter form of action can be considered schematic action (Doos et al., 2004).

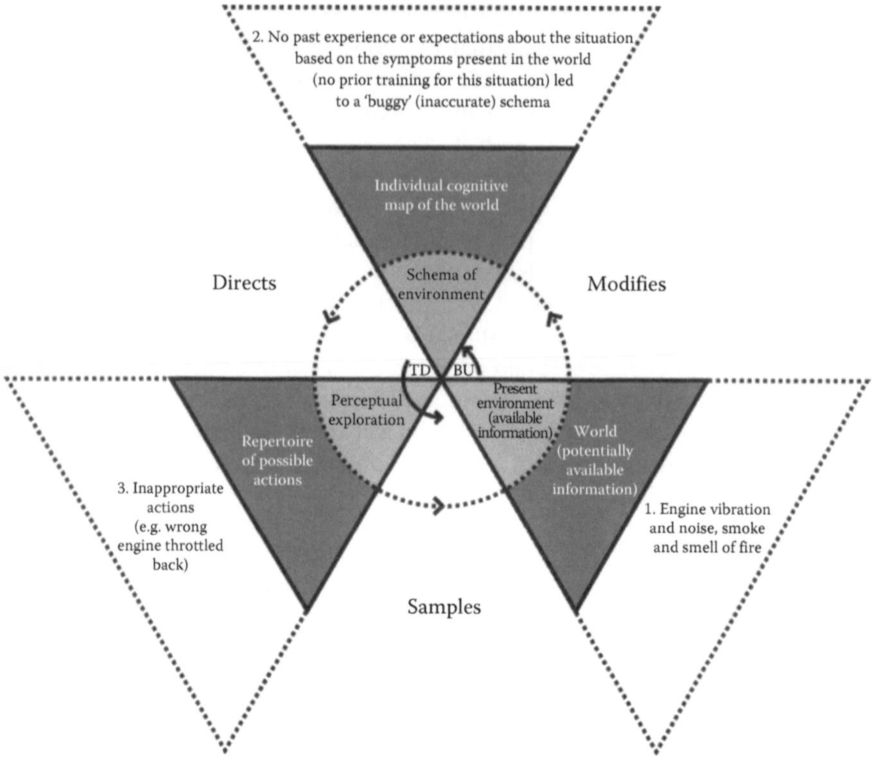

FIGURE 3.4 PCM for contributory factor 1.

Automatic activation is thus likely to lead to automatic action responses, hence a premature reaction to the problem. At the time of the Kegworth incident, the captain thought he had made the correct assessment of the situation and had no reason to suspect he had made any erroneous decisions. Furthermore, Kontogiannis and Malakis (2009) argued that problems with data interpretation are exacerbated when people either explain away or take immediate action (as occurred here) to counteract a symptom and later forget to integrate data that may be available (see contributory factor 3).

Norman and Shallice (1986) suggested that controlled processes (i.e. willed and guided attention) are only activated when either a task is too difficult (a novel situation) or errors are made. At this stage, the pilots were not aware that they were facing a novel situation, that is fan blade rupture rather than a more routine and trained for engine fire, nor were they aware that their decisions were to be erroneous. Therefore, a schematic perspective of the attention processes the pilots were engaged with would suggest they were automatic and thus relatively instantaneous based on their assessment of the situation which can account for the premature reaction to the situation, as summarised in Figure 3.5.

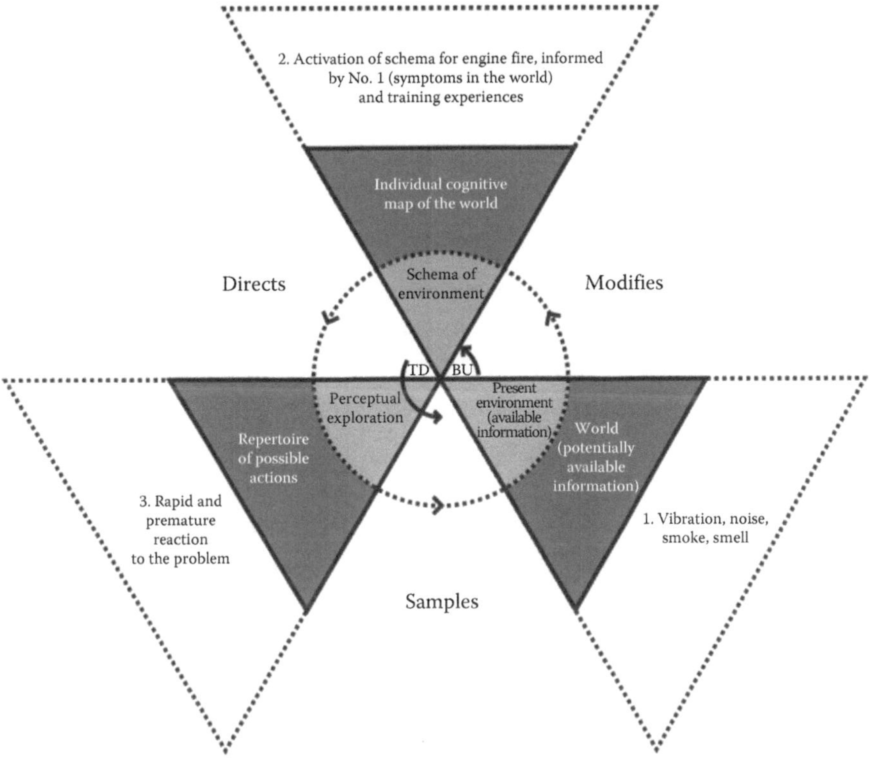

FIGURE 3.5 PCM for contributory factor 2.

3.3.4 Contributory Factor 3: Lack of Equipment Monitoring and Assimilation of Instrument Indications

> The engine parameters appeared to be stable, even though the vibration continued to show on the Flight Data Recorder (FDR) and were felt by passengers, but they were not perceived by the pilots. Additionally, the crew looking at the engine instruments did not get an indication of which engine was faulty. Furthermore, the engine vibration gauges that are part of the Engine Instrument System (EIS) were not included in the captain's visual scan as in the previous models of aircraft, it was considered unreliable.

One of the defining features of schemata is their emphasis on the role of past experience in guiding the way people perceive and act in the world. After the Kegworth accident, the captain stated that he rarely included the engine vibration gauges in his visual scans as he believed them to be unreliable and prone to false readings; this belief was based on his experience of these instruments in other aircraft. Therefore, his 'scanning schema' would not have included these instruments once

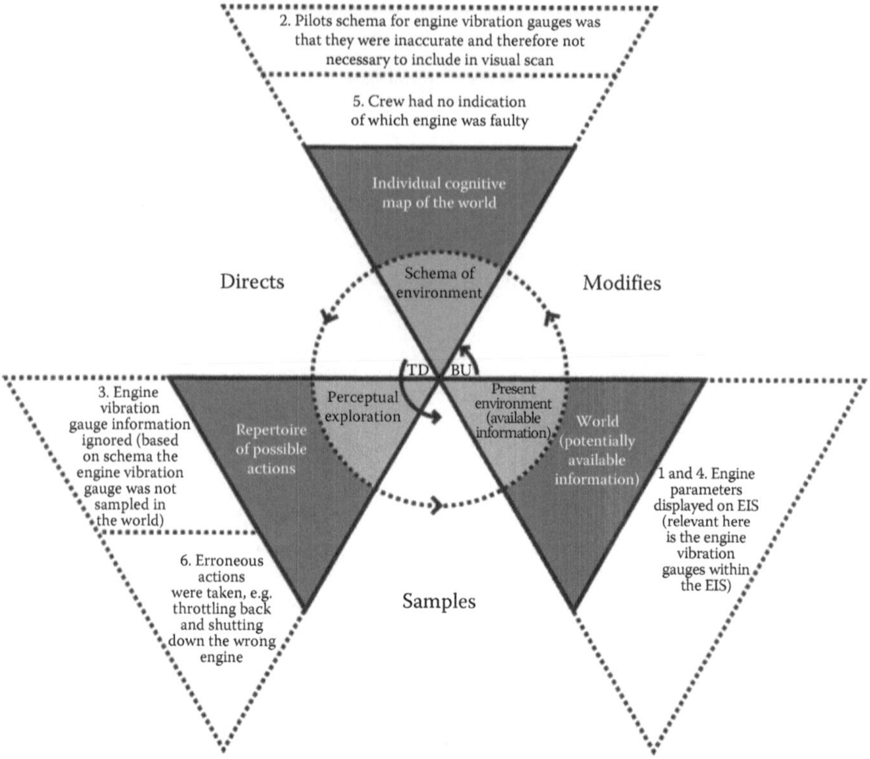

FIGURE 3.6 PCM for contributory factor 3.

it was activated. The active schema available to the captain was not based on the current model of the aircraft as he only had 23 hours flying experience in it. The prevailing and enduring view of the unreliability of the engine vibration gauges was formed from over 13,000 hours of flying experience with other aircraft types. As a result, the captain was left with a faulty schematic representation for the scanning process which is represented in Figure 3.6. Stanton and Walker (2011) discussed how expectancy can override any external cues to the contrary, and similarly, Rafferty et al. (2012, 2013) demonstrated the role of expectations as being an influencing factor in fratricide (friendly fire) situations in the military.

This contributory factor highlights how systemic factors can play a vital role in triggering error. Although the new engine vibration displays (digital rather than mechanical pointers) were technically more reliable, 64% of British Midland Airway pilots thought the new display system was not effective at drawing attention to changes in engine parameters and 74% preferred the old mechanical pointers (AAIB, 1990). It would appear that airline training failed to demonstrate how the new engine vibration gauges were more accurate. Training on the new EIS was included in the one day course that explained the differences between the Series 300 and 400 aircraft. There were no flight simulators equipped with the new EIS at the time; therefore, the first time a pilot was likely to see abnormal indications on the instruments was in-flight with a failing engine (AAIB, 1990). In NDM research, Klein (1998) terms missing events as negative cues; these are usually a stumbling block for novices, because the experience possessed by experts allows them to form and use expectancies. When these expectancies are misleading, however, the expert may still fall foul of missing events, in this case not assimilating the information from the engine vibration gauges. The AAIB report (1990) recommended that the Civil Aviation Authority (CAA) review their training procedures to ensure crews are provided with EIS display familiarisation in a simulator to acquire the visual and interpretive skills necessary; in other words, develop schemata for a range of failures and their representation on an EIS. These systemic influences would have played a part in creating the schemata the pilots relied on and therefore influenced their actions.

3.3.5 Contributory Factor 4: Cessation of Smoke, Noise and Vibration When the Engine Was Throttled Back

> Believing that the No 2 (right) engine had suffered damage, it was throttled back and eventually shut down. As soon as this happened the shuddering ceased and there was a cessation of the smoke and fumes, even though it would have been coming from the left engine. This 'chance' occurrence (that the left engine ceased to surge as the right one was throttled back) caused the crew to believe that their action had the correct impact.

People tend to assume that their version of the world is correct whenever events happen in accordance with their expectations, which has led to instances where schemata have been implicated in the cause of accidents (Besnard et al., 2004). For example, inconsistent schemata for the situation was said to have contributed to the actions that led to the Chernobyl disaster (Reason, 1990; Richardson and Ball, 2009). This phenomenon is termed 'confirmation bias' (Klayman and Ha, 1987) and refers to people seeking information that is likely to confirm their expectations. Besnard et al. (2004) applied, to the Kegworth incident, the conditions in which operators believe they have a good picture of a system or situation even though the underlying causal mechanisms have not been captured, such that two sequential events can happen as expected without capturing the cause and it is then treated as evidence to back up their assertion that their expectation was correct. The role of expectations in being a contributory factor in accidents is further supported by Malaterre (1986) who found that nearly two-thirds of driving accidents are the result of inappropriate expectations or interpretations of the environment. In the Kegworth example, the pilots believed they had a good picture of the system (i.e. their *schema* was that the right engine was damaged) and the two sequential events that happened were throttling back the right engine (*action*) and the cessation of vibration and fumes from the left engine (in *world*), convincing the crew that their assertion was correct. The reduction in the level of symptoms (engine vibration and fumes) lasted for 20 minutes and was compatible with the pilot's expectations of the outcome of throttling back the No. 2 engine; therefore, it is clear how this would have been taken as evidence that

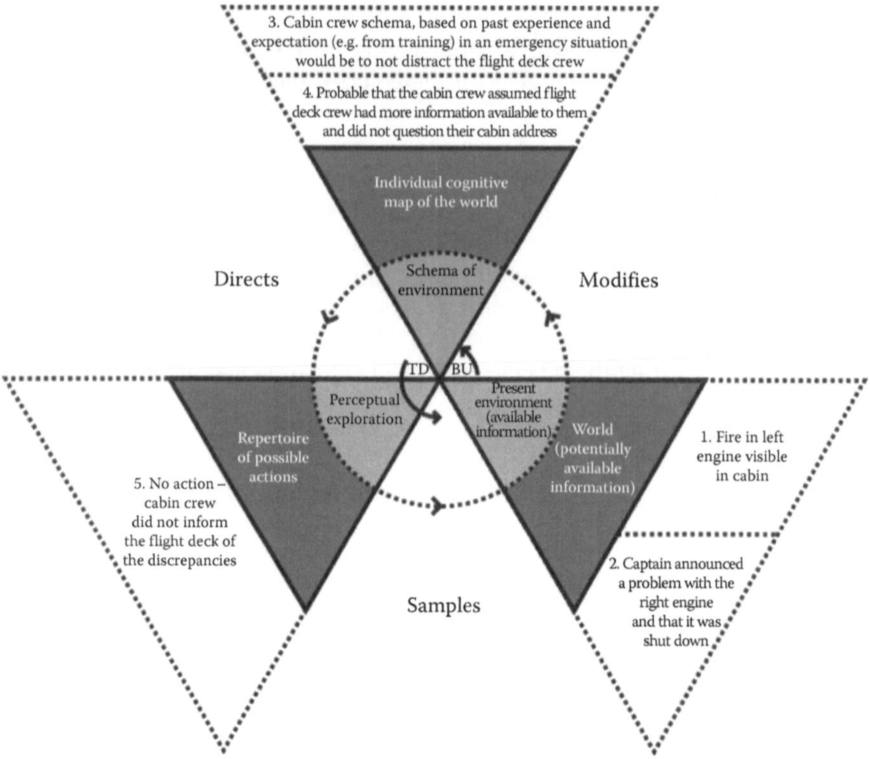

FIGURE 3.7 PCM for contributory factor 5.

the correct action had been performed. Figure 3.7 represents this contributory factor in the PCM.

This situation can be further explained by the first of the three categories of action slips that Norman (1981) refers to as 'slips during the formation of an intention'. This type of slip results from a situation being falsely classified (e.g. right not left engine was damaged). The action (throttling back the right engine) is intended and appropriate for the misclassified situation but not the actual situation (as the left engine was damaged). In the case of Kegworth, this confirmation bias is exacerbated by the lack of equipment monitoring by the pilots (see contributory factor 3) and the time-critical nature of the situation. As previously stated, Dekker (2006) argued that to understand error accurately it is essential to put oneself in the shoes of the operator and appreciate how the situation would have looked to them. The cessation of smoke and fumes led the crew to believe they had dealt with the problem and they can be forgiven for having thought this. It is only with the benefit of hindsight that one can appreciate how all the factors were interacting and how the situation should have been dealt with. A fuller explanation of what appears on the surface to be erroneous action can only be properly understood by putting the incident into context to consider the aetiology of events.

3.3.6 Contributory Factor 5: Lack of Communication from the Cabin Crew

> In the cabin, the passengers and the cabin attendants saw signs of fire from the left engine. The captain broadcast to the passengers that there was trouble with the right engine which had produced smoke and was shut down. Many of the passengers were puzzled by the captain's reference to the right engine, but none brought the discrepancy to the attention of the cabin crew, even though several were aware of the continuing vibration.

One of the key features of the reciprocal and cyclical nature of the PCM is how not only do schemata affect how people act in the world (i.e. direct action) but that information from the world can affect schemata. Therefore, had the pilots been informed of the confusion the passengers were experiencing due to the apparent misdiagnosis of smoke origin their schemata may have been revised accordingly. This contributory factor again emphasises how the PCM takes the system as a whole into account. Whilst the framework models perception and action of an individual (in this case, the pilots), the world in which the crew interact is the system element of the cycle. Systems are not a single entity but built up of various layers. Colleagues are part of these layers, and they can have an influencing effect on the perception and action of the operators at the sharp end. Woods et al. (2010) commented that operators are often unaware of the 'bugs' in their schemata and argued that feedback (potentially from people or machines) is one way to avoid this miscalibration of knowledge. Systems where feedback is poor are more likely to have miscalibrated operators, which appeared to be the case in the Kegworth incident. Figure 3.7 represents this contributory factor in the PCM.

The AAIB (1990) report suggested that the issue of 'role' influenced the passenger's acceptance of the captain's cabin address. Lay passengers generally assume the crew will have all of the information and knowledge available to them to have made an informed and correct decision. Similarly, the report suggested that although the cabin crew would have also been confused with the address they had no reason to suspect the pilots had not assimilated all of the engine parameter information available to them on the flight deck. In addition, cabin crew are aware that their presence on the flight deck could be distracting especially when the flight crew are dealing with an emergency. The AAIB report recommended joint training between flight and cabin crew to deal with such circumstances. This contributory factor, again, demonstrates how the wider system plays such an important part in the decision-making process and shows how an error, which initially looked to result from the actions of one pilot, are actually symptoms of trouble deeper within the system (Dekker, 2006).

3.4 DISCUSSION

In summary, the unforeseen combination of symptoms was outside of the pilots' training and experience. This led the pilots to base actions on what experience they did have, which has previously been shown to be the case with decision makers in complex situations (Morrison et al., 1997). The pilots' expectations about the engine vibration gauges meant they did not assimilate the readings on both engine vibration indicators. Additionally, the pilots' schematic representation of the air conditioning system was not appropriate for the unfamiliar aircraft type and contributed to the faulty diagnosis of the problem. Therefore, the actions that were taken were not appropriate for the situation, and the information that was available in the world (e.g. knowledge held by cabin crew or vibration gauge information) was not utilised by the pilots to update their schematic representations and modify their actions. From the thematic analysis of the AAIB report, it would appear that the PCM and Schema Theory offer a suitable theoretical foundation from which to explore decision-making. At the start of this chapter, five bullet points were presented that summarised the requirements of any 'good' account of error. Table 3.1 presents these requirements against how they have been accounted for

TABLE 3.1
Requirements of Human Error Research and How They Are Underpinned by Schema Theory and the PCM

Requirement from the Literature	Accounted for in the Kegworth Analysis
1. Human error needs to be the starting point of research and not the concluding label.	The errors made by the crew preceding the Kegworth accident were taken as the starting point for this research, and the reasons for these errors were investigated.
2. Error counting and classifying does not necessarily provide an understanding of why errors occurred nor is a causal account provided from this.	The use of Schema Theory and the PCM does not attempt to count errors. Whilst there is an element of descriptive classification (i.e. based on Norman's 1981 classification of error types cause by schemata), any classifications are used to further explain *why* an error occurred in a way that it did; for example, the PCM explanation accounts for why the EIS was not attended to.
3. Human error research needs to have a system perspective but still maintain a human-centred focus.	The PCM accounts for both of these factors by its interactive and cyclical explanation of perception and action. It is argued that perception and action in the world (system) are underpinned by operator's schemata. Similarly, the system can have a modifying effect on schemata.
4. Human error research must take the context of the work situation and the interaction between situation and mind into account.	The PCM accounts for interaction between mind and situation by its reciprocal and cyclical explanation of perception and action. The PCM provides an explanation that allow us to put actions and decisions in the context of the situation they occurred in.
5. The key question that any human error work needs to answer is *why* the actions and assessments made sense to the operators at the time.	This chapter answers the question of why the actions and assessments made sense to the Kegworth crew in the circumstances they were faced with. The PCM puts the operator in the context of the world they are working in and how both operator and world influence each other in a reciprocal and cyclical way.

by the Schema Theory and perceptual cycle model (PCM) approaches presented in this chapter.

Dekker (2006, p. 78) argued:

> …the only real explanation of error is that all factors come together.

Similarly, Woods et al. (2010, p. 8) contended:

> There is no single cause of failure rather a dynamic interplay of multiple contributors.

Both of these assertions were certainly the case in the Kegworth incident, without one of the factors (including lack of engine vibration gauge monitoring, assumed knowledge of air system and lack of crew communication) the accident may not have happened. The PCM and Schema Theory can account for these factors and how they influenced the perceptions and actions of the pilots to show how their actions made sense to them at the time.

3.4.1 Practical Applications

Schema Theory has much logical support (Schmidt, 2003), and many empirical studies support the notion of schemata in everyday life (e.g. see Brewer and Dupree, 1983; Hannigan and Tippens Reunitz, 2001). In aviation accident analysis, although not always explicitly stated, the notions of schemata are evident in many explanations of collisions (e.g. the 1998 Swissair crash in which the pilots did not effectively deal with smoke which then escalated into a full fire because their perception was that the air conditioning smoke was not immediately threatening). Furthermore, Rodrigues de Carvalho et al. (2009) reviewed the 2006 Amazonian mid-air collision and attributed some of the pilot errors to their use of schemata during their decision-making process. The pilots did not query ATC in relation to conflicting sets of flight plans because they used their general knowledge and past experiences (their schema for this situation), and it resulted in them rationalising that it has happened before. One of the pilots stated that having a flight plan for one altitude and being authorised to fly at another altitude happened 99% of the time, and so utilising this schema for the situation caused them to pursue the flight with conflicting information which had disastrous consequences. Reason (1990) argued that it is the interplay between psychological and situational factors that governs human behaviour; viewing decision-making through the perspective of the PCM enables both these factors to be taken into account.

The use of Schema Theory to understand error may lead to clear design and training implications (Ponomarenko, 2004; Zhang et al., 2010). Taking schemata into consideration is important when designing systems because human thinking involves the application of previous knowledge and expectations (Smith-Jackson and Wogalter, 2007). Klein (2008) criticised traditional training techniques that sought to teach rules and procedures as these are not applicable in dynamic sociotechnical systems like cockpits. Klein further argued that whilst systems approaches to training were an improvement, they still did not go far enough at increasing expertise or allowing for better decisions to be made. Similarly, Stanton et al. (2010)

acknowledged the fact that training systems should be designed to ensure the most efficient decisions are made. If, as argued here, the PCM provides an understanding of why people make certain decisions and what influences these judgements, then expertise can become the guidepost for training. Approaches to training and design need to focus on triggering appropriate schemata for skilled pilots. Whilst schemata are unique to individuals, Dekker (2006) argued that what made sense for one person will probably make sense for someone else; therefore, there seems merit in understanding individual perceptual processes and their role in decision-making. Similarly, Baksteen (1995) argued that the first level of underlying causes (of accidents) lie within psychological precursors of the crew, which will be common to mankind. Whilst these cannot be changed, they can be taken into account when designing procedures and systems; for example, Smith-Jackson and Wogalter (2007) found that designs in healthcare based on common schemata increased compatibility and facilitated usability.

Whilst there are many advocates of Schema Theory, the perspective never reached the prevalence in the literature that other theories have, such as the information processing approach, nor do some people, even champions of the approach, regard it as a true theory even though it appears to be a lot more logically intuitive than other competing approaches. However, as argued in Chapter 2, Schema Theory and the PCM do fit the generic requirements of a theory, and this analysis of the Kegworth accident has shown how the theoretical frameworks can provide this explanation.

3.5 CONCLUSIONS

This chapter applied the principles of Schema Theory and the PCM to the analysis of decision-making that resulted in the Kegworth aviation accident. From the case study analysis, it would appear that the PCM provides a promising avenue to explore the process of decision-making in the aviation domain. This work has contributed to the understanding of why the actions and assessments performed by the Kegworth pilots made sense to them at the time. The PCM provides a compelling explanation to account for system error, one that is particularly relevant in the aviation domain as a human-centred systems approach is provided. Chapters 2 and 3 have demonstrated that the strengths of Schema Theory and the PCM outweigh any criticisms levelled against them, and therefore, these ideas will be taken forward into subsequent chapters. Now that the theoretical principles have been identified, a method is required to explore aeronautical decision-making in more detail; this will be addressed in the next chapter.

4 A Pilot Study *Using the Perceptual Cycle Model and Critical Decision Method to Understand Decision-Making Processes in the Cockpit*

4.1 INTRODUCTION

The purpose of this chapter is to introduce the perceptual cycle model (PCM; Neisser, 1976), and its incorporation of Schema Theory, as a way of examining the process of aeronautical decision-making. A qualitative method is presented in which the PCM is used as a coding scheme to analyse data elicited from the critical decision method (CDM; Klein et al., 1989). In the previous chapter, the PCM was presented as a way to structure accident report data in order to infer explanations for decision-making. This chapter seeks to explore whether the PCM can be applied to interpret CDM data and how reliable this approach is. After introducing the relevant theoretical and methodological perspectives, the remainder of the chapter is split into three parts: Section 4.3 analyses a critical incident in relation to the PCM, Section 4.4 addresses the reliability of this method and Section 4.5 provides a general discussion of the practical implications and applications of this approach.

4.1.1 Aeronautical Decision-Making

Aeronautical decision-making (ADM) is a form of naturalistic decision-making (NDM; Klein et al., 1989) in which decision makers have domain expertise and make decisions in contexts which are usually characterised by limited time, goal conflicts and dynamic conditions. ADM is an important area of research as McFadden and Towell (1999) have argued that pilot judgement in decision-making and the handling of emergency situations is usually the deciding factor as to whether an incident will become an accident. Decisional errors have consistently been found to account for a high proportion of pilot error (Diehl, 1991; Orasanu and Martin, 1998; Shappell and Wiegmann, 2009). However, NDM is complex because there is often no clear standard of correctness, and there is a loose coupling between event outcome and decision process, so that outcomes cannot be used as reliable indicators for the quality of the decision (Orasanu and Martin, 1998). For example, good decisions may be overwhelmed by events outside of the decision maker's control, resulting in a bad outcome (Stanton et al., 2010). As exemplified in Chapter 3, with hindsight, it is easy to conclude that a poor decision was made. When examining NDM, it is therefore important to focus on the adequacy of the decision-making *process* rather than using the outcome as a benchmark for 'decision correctness'. To do this, it is necessary to understand how and why the actions and assessments made by an operator made sense to them at the time (Dekker, 2006). The cognitive origins and processes that underlie decision-making have traditionally been overlooked (Maurino, 2000; Hobbs and Williamson, 2002). It is proposed that Neisser's (1976) PCM is a suitable framework to model the cognitive basis of decision-making and with that comes the ability to identify areas vulnerable to potential mistakes and establish ways to mitigate these before errors occur.

4.1.2 Perceptual Cycle Model

To recap from Chapter 3, the PCM (Neisser, 1976) is based upon the idea of a reciprocal, cyclical, relationship between operator and environment. The PCM models interaction between person and world, with heavy emphasis on the role of schemata. As illustrated in Figure 3.1, Neisser presented the view that human thought is closely coupled with a person's interaction in the world, both informing each other in a reciprocal, cyclical relationship. World knowledge (schemata) leads to the anticipation of certain types of information; this then directs behaviour (action) to seek out certain types of information and provides a way of interpreting that information. The environmental experience (world) can result in the modification and updating of cognitive schemata, and this in turn influences further interaction with the environment.

Smith and Hancock (1995) have argued that the usefulness of the PCM explanation lies in the interaction between operator and environment rather than considering the two separately. They used the PCM as the foundation for their perspective of situation awareness (SA), arguing that it is impossible to comprehend SA without an understanding of the interaction between operators and their task environment. The same view can be taken when comprehending decision-making; without an explanation of the interaction between the operator and their work context, an understanding of why decisions were made is not achievable (Dekker, 2006). This chapter describes a method that has been employed to unearth the perceptual cycle process of decision-making when dealing with a critical incident.

4.1.3 Schema Theory and NDM

Schemata are described in detail in Chapters 2 and 3, but Schema Theory is discussed here in the context of NDM. Schemata are conceptualised as having 'slots' which are used to structure the information linked to them. Although schemata represent abstract concepts, they are built from specific instances and allow abstract knowledge to be derived at the time of retrieval by sampling from domain-specific instances. As described in the PCM, when an environmental experience is encountered, relevant past experiences (schemata) are retrieved to help develop an appropriate response.

There is general consensus that NDM is strongly schema driven (Klein, 1993). For example, Elliott (2005) has argued that schemata form the lenses through which the decision maker views the problem. Regardless of the domain, decision makers will use the schemata they possess to make decisions. Often the use of schemata in decision-making is advantageous; they act as natural standard operating procedures (SOPs) to direct decision makers to make appropriate responses to environmental stimuli based on previously successful past experiences. Problems arise, however, when the activated schema is inappropriate for the current situation (exemplified in Norman's (1981) application of Schema Theory to error). The role of schemata in decision-making research has been widely demonstrated, in both 'normal' situations (e.g. Walker et al., 2011) and erroneous situations (e.g. Stanton and Walker, 2011; Plant and Stanton, 2012a; Chapter 3).

Studying the role of schemata in decision-making comes from the perspective of the individual level of analysis; that is, schemata are unique individual cognitive structures. In Ergonomics research, there has been notable criticism about the reductionist nature of the individual level of analysis when compared to more systemic levels of analysis (Reason, 2000). However, it is argued here that whilst human operators still operate at the 'sharp end' of a system, there is still a need to understand the cognitive processes that underpin decision-making. The PCM places the operator (with their individual schema) into the environment in which the decisions take place and in doing so recognises that cognition is distributed within a system (Rafferty et al., 2010). As such, the PCM accounts for the operator–environment interactions that occur. As highlighted by Orasanu and Martin (1998), there is a loose coupling between event outcome and decision process. Furthermore, it is widely acknowledged that people make decisions according to the principle of local rationality; that is, behaviour (which in hindsight appears erroneous) is rational when viewed from the perspective of the current work context (Reason, 1990; Dekker, 2006).

Schema Theory, however, is not without its criticisms (see Chapter 2). One of the notable issues with schemata is how to elicit them. Plant and Stanton et al. (2013; Chapter 2) have acknowledged that the internal nature of schemata means they cannot be directly measured, but only inferred through the empirical consequences of them and qualitative methods are likely to be suitable for this. The difficulties of eliciting and representing schemata have previously been a barrier to the application and utility of Schema Theory. From the argument presented, it would appear that there is value in exploring the principles of the PCM, in which Schema Theory is a central tenet, as a way to structure and understand the process of ADM. The CDM has been utilised here as the qualitative data collection method in order to explore the role of the PCM in ADM. An explanation of the CDM follows.

4.1.4 Critical Decision Method

4.1.4.1 Description

The CDM (Klein et al., 1989) is one of the most commonly used cognitive task analysis methods (Stanton et al., 2005). The CDM achieves knowledge elicitation through the use of cognitive probes as a tool for reflecting on strategies and reasons for decisions during non-routine situations. Since its development, the CDM has been extensively used in a variety of domains including emergency dispatch management (Wong et al., 1997), critical care nursing (Crandall and Gretchell-Reiter, 1993), aviation (O'Hare et al., 1998) and transport (Goode et al., 2014; Young et al., 2015). Specifically, this method focuses on eliciting knowledge for behaviours Klein et al. (1986) classed as 'recognition-primed decisions', that is decisions for which alternative actions are derived from a recognition of critical information from the environment and prior knowledge. The emphasis on both prior knowledge and information from the environment makes the CDM a potentially suitable method to elicit data that can be analysed from the perspective of the PCM (which also places emphasis on prior knowledge, i.e. schemata, and the environment, i.e. world information). Furthermore, the CDM was selected as the data collection method because it is a theory-driven approach based on the assumption that expertise emerges most clearly during non-routine events (Klein et al., 1989). The Schema Theory literature suggests that the use of schemata is most evident in experts as opposed to novices, as experts have more past experiences from which to assimilate schemata (Elio and Scharf, 1990).

4.1.4.2 Procedure

The CDM achieves knowledge elicitation by asking people to discuss previous incidents they were involved with. The process of eliciting information is via cognitive probes in a retrospective semi-structured interview. Crandall et al. (2006) described the four phases for conducting a CDM interview:

1. *Incident identification*: The interviewee identifies several candidate incidents, and with the help of the interviewer, an appropriate incident is selected for deepening. The selected incident will depend on the goals for data collection. For example, if the research question is interested in understanding the role of schemata in decision-making, then incidents in which knowledge and experience affected the outcome of an incident are favoured. The interviewee is required to provide a brief account of the story from beginning to end.
2. *Timeline construction and verification*: This phase is aimed at getting a clear and refined overview of the incident structure by identifying key events that occurred in order to expand the initial account. This structure provides a framework for the remainder of the interview. Diagramming the timeline can be a useful exercise as often mistakes or gaps are identified and additional detail can be added. This exercise can also identify crucial decision points in which the interviewee experienced a major shift in their understanding of the situation or took action that affected the events.
3. *Deepening probes*: This phase is the most challenging as it allows an opportunity to get inside the head of the expert. This phase goes beyond the time elements and basic facts and attempts to understand the perceptions, expectations, goals and judgements of the expert. This is where the CDM probes are used to deepen the understanding of the event. Following the

advice in Stanton et al. (2005), if the participant gives a 'yes' or 'no' answer, prompts such as 'why' should be used to expand answers. A full list of the CDM probes is provided in Table 4.1. Probes should be defined prior to the analysis to ensure the output is compliant with the aims of the study (i.e. eliciting data that could help determine the role of schemata during critical decision-making). It is acknowledged that the probes have been modified over the years and researchers are encouraged to modify or add to the list as necessary for their individual research projects (Klein and Armstrong, 2005; Crandall et al., 2006). In light of the research goals here, five additional probes were added that expanded certain areas of the original CDM probes to increase their relevance to the elicitation of schemata (i.e. drawing out role of experience and expectations). These additional probes are italicised in Table 4.1. In the full version of the CDM, the probes are asked for each individual event identified in the timeline. A shorter version of the CDM can also be conducted in which the probes are asked only once, in relation to the whole incident (as opposed to each individual segment), this is useful if time is short.

4. '*What if' queries*: The final stage of the CDM interview provides an opportunity to consolidate the interviewer's insight into the interviewee's experience, skill and knowledge. The incident is used as a starting point for the interviewer to pose various hypothetical scenarios (about the overall incident or individual segments) in order to establish how the outcome of the incident may have been altered. Based on time constraints, Crandall et al. (2006) have acknowledged that this stage is not always essential to achieve project goals and as such can be omitted in a shorter CDM interview.

4.1.4.3 Evaluation

The main limitation with the CDM is its reliance on verbal reports (Klein et al., 1989). How far a verbal report accurately represents the cognitive processes of the decision maker is questionable (Stanton et al., 2005). People can misrepresent their own decision-making strategies and goals, especially as this is reported retrospectively and there are known issues associated with memory alteration and decay (Klein and Armstrong, 2005). Klein et al. (1989) argued that it is essential to acknowledge the biases associated with a method of this nature and work to reduce them. As such, to increase confidence in the reliability of this method a retest reliability study was undertaken. The result of this is reported in Section 4.4. This approach, however, is not always practical, and therefore, inter-rater reliability is usually calculated in relation to coding schemes used to analyse CDM data. For example, Klein et al. (1989) assessed the inter-rater reliability of the method in terms of how reliably a decision point could be identified from the unstructured portion of the interview. Percentage agreement between the two coders was over 80%, suggesting that decision points can be reliably identified with CDM data. Section 4.4 describes an inter-rater reliability study undertaken to assess the reliability of the PCM coding scheme used to analyse the CDM data in this study. Whilst there are issues associated with the CDM, the benefits are often seen to outweigh them (Klein et al., 1989). For example, the retrospective nature of the method means that the events of interest have occurred so there

TABLE 4.1
CDM Deepening Probes (Probes in Italics Were Added for This Research)

CDM Deepening Probes	
Cues	What were you seeing, hearing, smelling and noticing?
	For each phase, detail any physical events (e.g. alarm sounding) that defined the phase.
	For each phase, detail the mental events (thoughts, perceptions) that defined the phase.
Information	What information did you use when making the decision or judgement?
	How and where did you get this information, and from whom?
	What did you do with the information?
	At any stage were you uncertain about the reliability or relevance of the information that you had available to you?
	Was there any additional information you might have liked to assist with formulating the decision?
Analogy	Where you reminded of any previous experience?
	What was it about the previous experience that seemed relevant for this case?
Standard operating procedures	Does this case fit a standard or typical scenario?
	Is it a type of event that you are trained to deal with?
Goals and priorities	What were your specific goals and objectives during the incident?
	What was most important to accomplish at this point in the incident?
Options	What other courses of action were considered or were available to you?
	How was this option chosen and the others rejected?
	Was there a rule that you were following in choosing this option?
Experience	What specific training or experience was necessary or helpful in making this decision?
	Was the decision you made comfortable within your experience?
	Did your experience influence the decision that you made?
Expectations	Were you expecting this sort of incident to arise during the course of the flight?
	Were you expecting to make this sort of decision during the course of the event?
	Did your expectation influence your decision-making process, if so, how?
Mental models	Did you imagine the possible consequences of this action?
	Did you create some sort of picture in your head?
	Did you imagine the events and how they would unfold?
	At any stage, did you find it difficult to process and integrate the information available to you? (If yes, how did you overcome this?)
Decision-making	*What was the primary decision that you made?*
	What features were you looking for when formulating your decision?
	How much time pressure was involved in making the decision?
	How long did it take you to actually make this decision?
	At any stage, were you uncertain about the appropriateness of your decision?
Guidance	Did you seek any guidance at this point in the incident?
	How did you know to trust the guidance you got?

is no need to create artificial simulations that are limited in contextual richness, nor is there a need to wait for non-routine events to occur (Klein et al., 1989). Klein et al. (1989) have argued that the semi-structured nature of the interview results in less time to gather relevant information but retains the freedom to explore interesting data. The standardised probes also result in more reliable data.

4.2 METHOD

4.2.1 Methodological Perspective

The main impetus for this chapter was to explore a method that allows the role of the perceptual cycle in decision-making to be understood. As previously discussed, it is essential to understand the decision-making *process* rather than just the decision outcome and the PCM and associated Schema Theory appear to be useful theoretical perspectives from which to do this. The issue of inferring schemata has been raised and the selection of the CDM justified.

The method undertaken was based upon the principles of thematic analysis by which text data are classified into meaningful themes. Here, deductive thematic analysis was conducted, in which themes (patterns in the data) are generated from existing theory (the PCM) (Boyatzis, 1998). In accordance with the objectives of this chapter, the coding scheme was based on the categories of the PCM and developed in line with Boyatzis' (1998) five criteria of how to structure a meaningful code.

Table 4.2 provides the criteria along with the coding scheme used in this analysis. This form of qualitative data analysis is open to criticism about the reliability of the results. As Klein et al. (1989) acknowledged, the CDM is a qualitative data gathering technique and such data cannot escape subjective interpretation. Assessing the reliability of the data interpretation, however, can go some way to making the claims more objectively verifiable. Both inter- and intra-rater reliability are critical in thematic analysis, where computing it combines the richness of qualitative information with the precision of quantitative methods (Boyatzis, 1998). Singletary (1993, p. 294) argued:

> ...if the coding is not reliable, the analysis cannot be trusted.

4.2.2 Study Participant and Procedure

The CDM procedure outlined in Section 4.1.4.2 was employed with a search and rescue helicopter pilot. The first interview (T1) occurred 9 months after the critical incident (T0). The second CDM interview (T2, conducted to assess the test–retest reliability of the CDM) took place 25 months after T1. The participant was a male helicopter pilot (aged 34 years at T1). He was voluntarily recruited through an advert for participants placed on the British Helicopter Association website. At T1 and T2, a face-to-face interview was conducted which lasted for approximately one hour. At T1, the participant had approximately 2500 flying hours. The incident occurred when flying an AW139, at T0 the pilot had approximately 300 hours on this type of aircraft. At the time of the incident, the helicopter was being flown for a search and rescue training exercise. This is a domain that has recently been used by other researchers, such as Baber et al. (2013), as it eloquently exemplifies the NDM

TABLE 4.2
Criteria for a Meaningful Coding Scheme (Boyatzis, 1998) and the PCM Coding Scheme That Was Developed

Criteria for a Meaningful Code (Boyatzis, 1998)	**Coding Scheme Developed**		
1. A label/name	Schema	Action	World
2. Definition of what the theme concerns	Mental structures held by individuals that organise their representations of the world. Schemata are heavily influenced by past experiences and expectations.	The process or statement of doing something, or the intention to do something.	Externally available information in the world (environment).
3. Description of how to know when the theme occurs	Statements relating to the use of prior knowledge and experience, i.e. things based on experience, expectation or 'knowing' things (this could be implied information through the discussion of training and/or standard operating procedures).	Statements of doing an action or discussion about potential actions that could be taken (mental or physical actions).	Statements relating to potential or actual information existing in the world (environment). Can be physical things, conditions or states of being.
4. Description of any exclusions	References to mental information made in the context of an action should be coded as action. For example, *talking about known training procedures* should be coded as action, even though the statement refers to prior knowledge of training procedures.	Only code as action if the statement is referring to explicit actions made by the pilot/crew. For example, the statement *light came on* is an action, but in this context this statement provides the pilot with information about the state of something in the world (i.e. light is on) and would therefore be coded as 'world'. The statement *I turned the light on* is coded as 'action'.	n/a
5. Example	*My expectation was that the engine would take a while to start in the rain.* *…knowledge of training helped…* *…I had no other experience to base it on.*	*I turned on the engine.* *…we'll turn it off and on again…* *…Cancel the training and head back to base…*	*Caution light came on.* *It was raining.* *…the primary flight display screen was on…*

environment in a rotary wing context. Ethical permission for this and the subsequent studies reported in this chapter were granted by the Research Ethics Committee at the University of Southampton.

The nature of the interview was explained to the participant; that is, he would be required to recall an incident and the interview would last approximately 1 hour depending on length of answers provided. A critical incident was defined as being 'a non-routine or un-expected event that was highly challenging and involved a high workload'. Due to time constraints, the shorter version of the CDM was conducted. The CDM interviews were audio recorded and transcribed. In accordance with guidelines on qualitative data analysis, text was chunked into meaningful segments of approximately one sentence or less in length (Strauss and Corbin, 1990).

4.2.3 Methodological Questions to Answer

The theoretical arguments that have been presented have raised several methodological questions that this chapter will seek to answer:

1. Is the PCM able to account for the decision-making process of a pilot dealing with a critical incident using data collected from the CDM?
2. How reliable is the CDM in terms of test–retest reliability?
3. How reliable is the method of qualitative data analysis in terms of inter- and intra-rater reliability?

4.3 INCIDENT ANALYSIS

4.3.1 Incident Synopsis

The pilot interviewed was the pilot not flying (PNF) on the day of the incident and was therefore responsible for inputting data into the aircraft systems and fault diagnosis. Other crew on board included the pilot flying, who had primary responsibility for the aircraft, and the rear winch crew. The PNF stated that the winch crew would have been aware of the situation via the intercom, but they were not involved in the diagnostic decision-making process. It should be acknowledged that the data presented in this chapter are from the perspective of the PNF as the level of analysis was defined at the individual. Stanton et al. (2010b) have discussed the debate that exists over establishing the appropriate level of analysis in the context of SA research. The authors have argued that as long as the level of analysis is defined and declared then research at the levels of the individual, the whole system and all the areas in between have a valid place in Ergonomics research. Due to this being an individual level of analysis, the roles of the additional crew members were not explicitly addressed. Issues associated with, what is likely to be, a composite account of the incident, are addressed in the discussion (Section 4.5). What follows is a synopsis of the incident amalgamated from the CDM interviews at T1 and T2:

> ...Finished winching in the English Channel and were then going to go to cliff winch... put in the waypoint, or what I thought was the waypoint, into the navigation system...I typed in the three digit code...then we looked and all the screens, well the 4 primary

screens, went blank. There was no navigation information and no primary flight information, except for the secondary back up system which gives an attitude, a speed and a height. It was a clear day with good conditions in uncontrolled airspace. Once we got over the initial shock...the immediate thought was *what's it done now?* blaming the systems rather than any action I made...by this stage we were expecting electronic problems, can guarantee one on most flights. We initially started looking for circuit breakers, to look if any had popped...lots of electronics on this aircraft, usually the case of finding the right circuit breaker. None had popped...went to the flight cards... there wasn't one for that, it is not the sort of thing that is expected...we tried turning up the brightness of the screens to see if that made a difference. Nothing else we could do...agreed we would head back to base to get the aircraft serviceable...headed north, we were in the Channel so if you head north you will hit the mainland somewhere, we knew once we picked up that we were okay. After the event we realised I had reverted to the old waypoint for the previous aircraft we flew. Shut the engine down and started it back up and the waypoint had cleared and the screens were back.

4.3.2 Thematic Analysis of Critical Incident

The coding scheme presented in Table 4.2 was used to analyse the data obtained from the CDM interviews. In line with the CDM procedure described in Section 4.1.4.2, a timeline of events was created and is presented in Figure 4.1. For each event in the timeline, the relevant data were coded with the PCM coding scheme to understand the perceptual cycle the pilot engaged in when dealing with the critical incident. Table 4.3 presents the data from each phase of the incident, organised by each element of the PCM. This is illustrated in Figure 4.2.

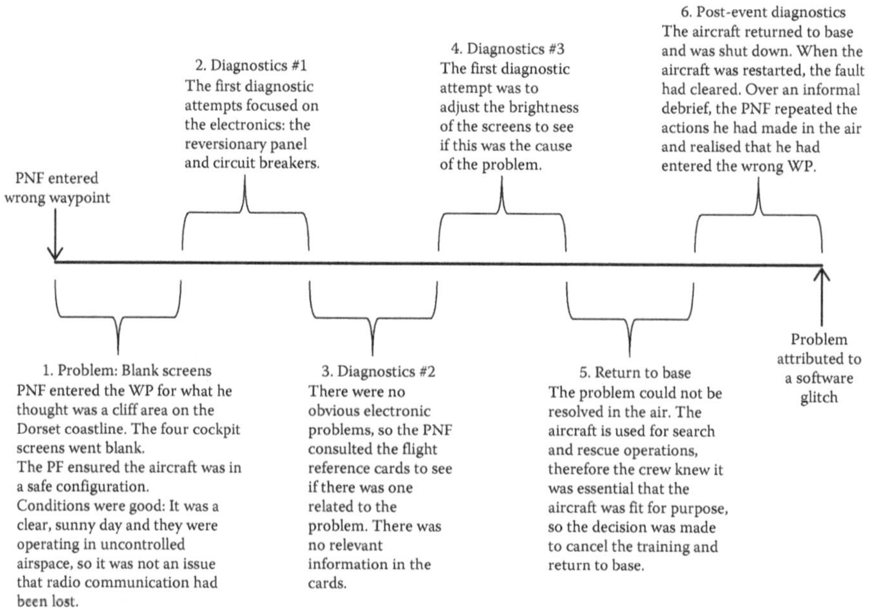

FIGURE 4.1 Timeline of the critical incident.

TABLE 4.3
Perceptual Cycle Process for Each Phase of the Critical Incident

Incident Phase	Code	Description
Onset of problem (blank screens)	W_1	During a winch training exercise, the PNF was required to enter a waypoint (WP) into the navigation system.
	S_2	PNF held a schema for the WP names developed from aircraft he had previously flown: *I reverted to the old WP for the S61...we have WP lists but was doing it from memory.*
	A_3	PNF keyed in the wrong WP (this resulted in the screens going blank).
	Misc.	The PF ensured the aircraft was in a safe configuration.
Diagnostics (electronics)	W_4	Blank screens, aircraft in safe configuration, AW139 has more electronics than previous aircraft.
	S_5	In his interview, the PNF stated, *We get a lot of electronic problems...can guarantee one on most flights, through our experience we understand which ones are important...aircraft often has little glitches.* This demonstrates that the PNF held a schema for the situation that the problem was likely to be related to the electronics: *...knew from ground school that there were two separate* [electronic] *systems, going to two separate screens.*
	A_6	*I played around with the* [reversionary] *panel to see if that made a difference, manually applied the feeds from the other side.*
	W_7	Screens were still blank.
	S_8	*The training we now do in the simulator is to check the circuit breakers.* The PNF held a schema developed from training experiences.
	A_9	Looked to see if any circuit breakers had popped.
Diagnostics (flight reference cards)	W_{10}	Screens still blank, not an immediately obvious electronic problem (e.g. no popped circuit breakers).
	S_{11}	Schemata acquired through training.
	A_{12}	Checked the flight reference cards to see if there was one relevant to this situation.
Diagnostics (screen brightness)	W_{13}	Screens still blank, no relevant flight reference cards.
	S_{14}	*We know that the screens* [brightness] *are sometimes turned down for a night flight...they would appear blank in daylight.* The pilot stated that this was highly unlikely to be the cause of the problem as the screens had been on previously, but he was exhausting all other options.
	A_{15}	*I tried playing with the switch* [for screen brightness] *to see if that made a difference.*
Decision (return to base and shut down)	W_{16}	Screens blank, diagnostic attempts had not located the problem, aircraft was flying fine, it was a clear day and they were in uncontrolled airspace.
	S_{17}	Both pilots were familiar with the location they were in, the PNF stated: *...in the English channel, south of the Isle of Wight, if head north I know we'll hit the mainland and be okay...almost certainly knew it was a glitch rather than a major system fault.*
	A_{18}	The PNF assisted the co-pilot when returning to base (e.g. eyes out navigation).
	Misc.	A collective decision was made to return to base, the aircraft was flown back by the pilot flying.

(*Continued*)

TABLE 4.3 (*Continued*)
Perceptual Cycle Process for Each Phase of the Critical Incident

Incident Phase	Code	Description
Post-event diagnostics	W_{19}	Back at base with the engineers.
	S_{20}	Talking about what had happened, *realised I put in the WP for what I believed to be Anvil Point, was used to flying the S61 with a different set of WPs...put in what I thought was the correct WP* (the WP had changed on the new aircraft).
	A_{21}	Told the engineers what I had done.
	Misc.	When the aircraft was shut down and powered back up, the problem cleared and the screens returned.

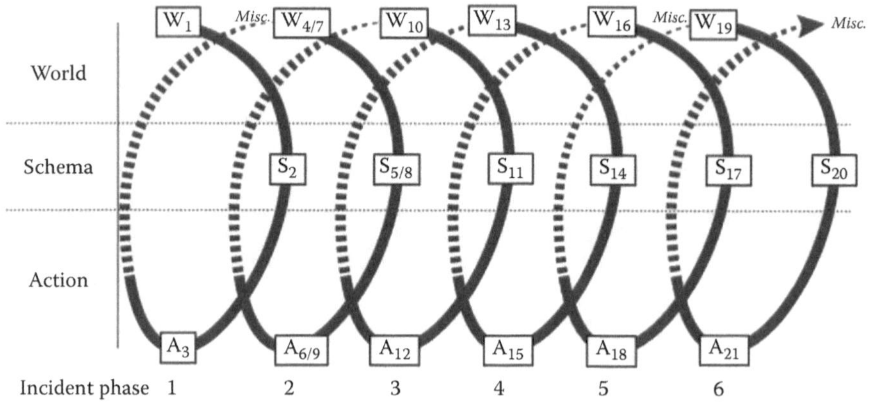

FIGURE 4.2 The progression of the incident through the perceptual cycle (descriptions for each annotation are listed in Table 4.3).

4.3.3 Incident Summary

The fundamental principle behind Neisser's (1976) PCM is that whilst it is structured, information does not move in a simple linear flow but rather a cyclical, interactive, process. This is illustrated by the helical representation in Figure 4.2. The schemata held by a person directs their actions and exploration in the world; that is, expectations lead to anticipating certain types of information in the environment which is actively sought out. Action requires information if it is to be carried out effectively, which in turn produces more information for the perceiver. The state of the world is then encompassed back into, and can have a modifying effect on, the individual's schemata and thus future interactions in the world. It is clear that in this incident the pilot's perspective was compounded by a host of issues that are associated with the schemata he held about the aircraft he was co-piloting. The pilot's assumption about the problem was based on his experience and expectations of the aircraft's electrical system, which led the pilot to engage with various diagnostic attempts in order to solve the problem of the blank screens. It is during this diagnostic phase that the pilot's reliance of his schemata for the situation is evident. A schema is an organised mental pattern of thoughts or behaviours to help organise world knowledge (Neisser, 1976). Decision-making literature suggests that decision makers make an initial assessment of a situation by looking for familiar patterns or prototypes (Klein et al., 1986). Amongst other things, this strategy saves cognitive resources by generating appropriate options and responses. Based on the pilot's experience of the electrical glitches often encountered in the aircraft, the pilot was expecting this fault to be of similar nature; that is, he was looking for familiar patterns.

The error made by the pilot of entering the wrong waypoint can also be explained by Schema Theory. Schemata are knowledge structures based on a set of similar past experiences, and they capture the common features of this experience. Minsky (1975) discussed frames as a form of static schema and argued that they have slots to be filled with information. These slots have a default value assumed; that is, the most common representation is used if alternative information is not provided. In the case of entering the wrong waypoint, the pilot reverted to the default value for his waypoint schema, that is the waypoint that was often used (correctly) in the previous aircraft but was incorrect for his current aircraft. Incidentally, the problem was caused by a software glitch. In the current aircraft, which was relatively new to the pilot, the entered waypoint was assigned to a location in New Zealand. The software glitch resulted in the screens being shut down if the waypoint location was over 10,000 miles away.

4.4 TESTS OF RELIABILITY

4.4.1 Reliability of the CDM

As detailed in Section 4.1.4.3, the reliability of the CDM has been questioned as the method relies on retrospective verbal reports; however, little appears in the literature assessing the retest reliability of the CDM. Taynor et al. (1987) conducted one of the few studies available to determine the retest reliability of the CDM. Data collected from eight firefighter commanders at time intervals of 3 days, 3 months and 5 months from when the incident was first recalled were compared. Correspondence of information across the different reporting times averaged 86%. In the present study, retest reliability was assessed with a time interval of over 2 years.

4.4.1.1 Procedure

To assess the retest reliability of the CDM, the procedure detailed in Sections 4.1.4.2 and 4.2.2 was employed on two different occasions. At T2, the participant was asked to recall the incident as if the researcher had never heard about it. The CDM interview at T1 was broken down into 135 text segments and the CDM interview at T2 consisted of 174 segments. The tabulated text segments were subjected to open coding, in which they were analysed to identify themes into which they could be grouped. Theme identification was assisted by the names of the phases provided by the participant in the timeline construction stage of the CDM interview. Constant comparison technique was employed, whereby each text segment was compared with previous items to see if the same or a different phenomenon was described. A second rater re-coded 20% of the text segments using the theme descriptions provided in Table 4.4.

4.4.1.2 Results

Percentage agreement between the criterion coder and inter-rater coder was calculated at 92%, well above the suggested threshold of 80% agreement (Jentsch and Bowers, 2005). T1 and T2 data were compared in terms of text segments in each theme. This is what Hoffman et al. (1998) described as a moderate check of reliability, that is that the general gist and details of an incident are the same, as opposed to an exact check of reliability whereby the same details are recalled in the same order at different time intervals. The point of this retest study was to assess the reliability of the CDM; a highly reliable method will produce similar data at T1 and T2. Table 4.4 presents key differences between data at T1 and T2.

4.4.1.3 Preliminary Discussion

The high level of inter-rater agreement (92%) between the criterion coder and inter-rater coder suggests that the themes used to classify the data were accurately representing what the data were describing. However, it is the data within each theme that is of interest in this retest study. Data obtained from the CDM interview at T1 and T2 were compared in terms of content within each theme. Table 4.4 details the themes identified in the data and where differences occurred. There were no differences between data at T1 and T2 in 9 of the 17 themes identified; some of the differences identified in the remaining 8 themes were marginal. Table 4.5 provides a more detailed description of the differences between the T1 and T2 data.

As expected, this retest study demonstrated that there are some differences in data elicited by the CDM at T1 and T2. Literature on human memory would argue that differences occur because memory alters and is distorted with the passage of time (Lieberman, 2012), and this would explain some of the differences that were found between T1 and T2 data. This is one of the primary concerns with using the CDM for data collection. Furthermore, Schema Theory suggests that schemata change and assimilate overtime based on recent world experiences and as such can have a modifying effect on subsequent actions and decisions, thus what is reported in a retrospective interview. As discussed in Table 4.5, this is evident in some of the differences found between the two data sets. However, considering the amount of time elapsed between the two CDM interviews (2 years), the data are generally very similar and the salient points of the incident were reflected in both interviews: the pilot discussed inputting the wrong waypoint, losing all four cockpit screens, various diagnostic attempts that were undertaken and the decision to abort the training and return to base. This retest study was conducted nearly 3 years after the incident; a reliability study with a time elapse of this length has not, to our knowledge, been conducted with CDM data previously. The overall conclusion that

TABLE 4.4

Theme Name, Description and Comparison of T1 and T2 Data to Assess the Retest Reliability of the CDM

Theme	Description of Theme	Key Differences between Data at T1 and T2
Conditions	Text relating to environmental conditions, such as weather, time of day	None
Aircraft status	Text relating to the current status of the aircraft in terms of performance and configuration	None
Waypoint (WP) information	Text relating to waypoints, e.g. inputting waypoint data and the associated navigation system, the role of waypoints and waypoint names	T1: St. Albans Head named as the WP T2: Anvil Point names as the WP and a more thorough description was given as to why the mistake was made (based on the use of the WP system)
Location information	Text relating to area information, e.g. location of aircraft, potential location of aircraft and location of other key landmarks	None
Communication information	Text discussing communications between aircraft crew and external operators such as the coastguard; includes text relation to the radio and other communication systems	T1: Theme not present T2: Data related to not being able to tune in the radios, who the crew could/couldn't communicate with
Problem	Text relating to the onset of the problem (blank screens)	None
Immediate actions	Text discussing immediate actions taken by crew at the onset of the problem	T1: The pilot said the aircraft was put into a safe configuration, i.e. the autopilot was put on T2: The pilot stated the autopilot was already on and aircraft was already in a safe configuration
Decision	Text relating to the decisions taken by the crew, e.g. return to base	None
Options	Text discussing alternative options available to the pilot, other than the primary decision that was made	None
Diagnostics (general)	Text relating to general diagnostic information, i.e. discussions about needing to understand what was wrong, as opposed to any specific diagnostic attempts	None

(Continued)

TABLE 4.4 (*Continued*)
Theme Name, Description and Comparison of T1 and T2 Data to Assess the Retest Reliability of the CDM

Theme	Description of Theme	Key Differences between Data at T1 and T2
Diagnostics (electronics)	Text about specific diagnostic attempts regarding the electronic systems	None
Diagnostics (flight reference cards)	Text about specific diagnostic attempts regarding the flight reference cards	T1: The pilot stated the crew looked in the flight reference cards T2: Theme not present (although it was alluded to in relation to general emergency checks)
Diagnostics (circuit breakers)	Text about specific diagnostic attempts regarding the circuit breakers	T1: A lot of text related to checking the circuit breakers T2: This was not explicitly stated; circuit breakers were discussed in relation to other electronic diagnostic attempts
Diagnostics (screen brightness)	Text about specific diagnostic attempts regarding screen brightness	T1: Theme not present T2: Screen brightness, i.e. adjusting the brightness to see if the screen had been turned down, was discussed as a diagnostic attempt
Post-event diagnostics	Text discussing what happened after the incident was contained (returned to base), i.e. post-event diagnostics and establishing the cause of the problem	T1: Theme not present (incident description ended with returning to base and shutting down the engine) T2: Post-event diagnostic information was provided relating to how the problem was solved
Similar incidents	Text referring to whether similar incidents have arisen since the incident and how these were dealt with	T1: Theme not present T2: The pilot discussed how the incident had occurred since and how it is now dealt with
SAR role	Text discussing the SAR role of the helicopter and how this influenced the decision-making process, e.g. choosing to terminate training and the importance of having a serviceable aircraft	None

TABLE 4.5
Differences between T1 and T2 CDM Data

Theme	T1 Data	T2 Data	Discussion/Implication
Waypoint location	St. Albans Head named as the waypoint	Anvil Point named as the waypoint	Both locations are cliff points on the Dorset coastline approximately five miles apart. In a post-interview debrief, the pilot was asked about the discrepancy. He stated that St. Albans Head was the correct location. The T1 interview took place 9 months after the incident, whereas the T2 interview occurred 34 months after the incident. It is unsurprising that nearly 3 years later the pilot confused cliff names; in the elapsed time, both locations were visited numerous times. This exemplifies that the retest study is seeking to establish what Hoffman et al. (1998) have described as 'moderate reliability', i.e. the general gist is the same (the training exercise occurred at a location on the Dorset coastline), but not necessarily presented in the same order or using exactly the same language.
Communication information	Theme not present	Pilot mentioned that due to the loss of screens they were not able to tune the radios, but the intercom still worked so they were able to talk to the rear crew	* The ascription of 'theme not present' only occurred in T1 data, i.e. everything mentioned at T1 was also mentioned at T2, but T2 also had additional information. In terms of text segments, there were more data in T2 (174 segments) compared to T1 (135 segments), so it is unsurprising that more themes are present in T2 data. Reasons for this are open to much speculation; for example, unbeknown to the researcher the pilot may have had more time available at T2 and therefore provided a more detailed interview. The process of having recalled the incident once, even though it was 2 years previously, may have also triggered subsequent memories and at T2 the pilot was more familiar with both the interview format and the researcher.
Diagnostics (screen brightness)	Theme not present	Adjusting the brightness to see if the screen had been turned down was discussed as a diagnostic attempt within the context of a more detailed discussion about the electronics of the aircraft at T2. For example, there was more mention of the way the reversionary panel worked	See *. ** Furthermore, this difference is potentially an example of the pilot being more familiar with the researcher. Due to resource constraints, the researcher conducting the interview was the same at T1 and T2. Although at T2 the pilot was instructed to recall the incident as if the researcher had never heard it, there may have been some unconscious processes at play by which the pilot felt he could go into more detail because 'the researcher had heard it before'. Furthermore, the pilot may have held an unconscious bias that after working in the field of human factors aviation research for a further 2 years, the researcher was more familiar with aviation terminology; hence, he went into more detail about the electronics.

(Continued)

TABLE 4.5 (*Continued*)
Differences between T1 and T2 CDM Data

Theme	T1 Data	T2 Data	Discussion/Implication
Similar incidents	Theme not present	The pilot discussed how the incident had occurred since and how it is now dealt with	See *. Furthermore, at T1 the pilot had not experienced the loss of the four screens before, at T2 it had happened to him on subsequent occasions, hence the 'similar incident' theme occurring in T2 data but not in T1 data.
Immediate actions	The pilot said the aircraft was put into a safe configuration, i.e. the autopilot was put on	The pilot stated the autopilot was already on and aircraft was already in a safe configuration	In terms of a moderate test of reliability, the data relating to immediate actions in the two data sets are relatively similar; the phrase 'safe configuration' and 'autopilot' was used in both interviews. The discrepancy lies in the order of events but this can be explained by the role of the pilot: at the time of the incident, the pilot was PNF, i.e. he was the co-pilot, responsible for tasks such as navigation and data input. The action of ensuring that the aircraft was in a safe configuration would have been performed by the PF, so it is unsurprising that a slight discrepancy arose in relation to actions that were not performed by the PNF.
Diagnostics (flight reference cards)	The pilot stated the crew looked in the flight reference cards	Less explicitly referenced, alluded to in relation to general emergency checks	A similar explanation to the screen brightness theme (**) can be applied here. Checking the flight reference cards (discussed in more detail at T1 than T2) is a fundamental part of dealing with a critical incident. The pilot may have held an assumption that it was not necessary to go into this in as much detail at T2.
Diagnostics (circuit breakers)	Much discussion about checking the circuit breakers	Less explicitly referenced, they were discussed in relation to other electronic diagnostic attempts	This can be explained in terms of Schema Theory; Schema Theory proposes that schemata are not static entities but instead dynamic and active. This is represented in the PCM, whereby world or environmental information can have a modifying effect on schemata. In the post-interview debrief, the pilot stated that training around the time of the incident (and T1) was extremely focused on the importance of checking circuit breakers; hence, it was fresh in his mind. This demonstrates how information in the world (e.g. recent training about circuit breakers) can have a modifying effect on the schemata held for a particular situation (e.g. checking circuit breakers to deal with the incident) and result in certain actions (e.g. identifying the diagnostic attempt of checking circuit breakers at the T1 interview).

has been drawn from this is that given such a considerable amount of time elapsed, and during that time the pilot would have dealt with a number of other incidents, recall was remarkably robust. This can be attributed to the use of the structured CDM probes, as they allow the same questions to be asked, and therefore, similar responses are elicited.

4.4.2 Reliability of the PCM Coding Scheme

Aside from establishing the reliability of the CDM, the reliability of the PCM coding scheme also needs to be considered. This is what Boyatzis (1998) termed 'rater-expert reliability', whereby a set of correct (or more correct) answers to the judgement situation have already been assigned (by an expert). Additional coders are judged against the standard set by the expert in a blind condition; that is, raters are unaware of the experts' coding decisions (Walker, 2005). Burla et al. (2008) argued that calculating inter-rater reliability is an established method to analyse the quality of a coding scheme. Here, inter-rater reliability of the PCM coding scheme was conducted on 65 text segments from the T1 CDM interview.

Reliability scores were calculated based on percentage agreement, that is the number of agreements divided by the number of times the coding was possible, multiplied by 100; this is in accordance with literature that has suggested this is the most suitable way to calculate reliability scores with data of this nature. For example, Boyatzis (1998) has argued that percentage agreement should be used when themes to be coded (in this case three) and numbers of observed situations (segments, in this case 65) are few. Three themes and 65 segments are relatively few in relation to other coding situations in which infinite themes can be identified from many hours of interview data. Furthermore, Boyatzis (1998) argued that if data resulting from the coding are nominal, percentage agreement scores are appropriate. Within the literature, there are conflicting accounts as to the acceptable level of percentage agreement as there are no established standards (Lombard et al., 2002; Marques and McCall, 2005). There is a general consensus that a level of 80% agreement and above indicates an acceptable level of reliability (Jentsch and Bowers, 2005); therefore, this will be the benchmark for assessing reliability in this study. Percentage agreement will also be used to assess the intra-rater reliability of the coding scheme; that is, the same codes are assigned by the same coder on two separate occasions.

4.4.2.1 Participants

A repeated-measures within-subjects design was employed in a study to assess inter and intra-rater reliability of the coding scheme presented in Section 4.2 (Table 4.2). The study required completion of a coding task conducted twice, exactly 4 weeks apart, with the same group of participants. Twenty postgraduate research students at the University of Southampton were voluntarily recruited from an e-mail advert. Seventy-five percent of the sample were male. The mean age of the participants was 27 years (range between 21 and 40 years, SD = 4.5). All participants rated English as either their first language (45% of sample) or as good as their first (55% of sample).

The participants were paid £20 for taking part in the study, which was paid on completion of the second coding task.

4.4.2.2 Materials

A presentation was given that included a description of the PCM with details of the coding scheme (Table 4.2) and six coding examples. The participants were provided with a coding guide that included a description of the PCM and examples of the coding scheme, a practice question/answer sheet and an answer booklet which included a page of demographic questions and the coding task. The coding task consisted of extracts of the incident transcript in continuous prose to allow the interview to be read and understood in context and then each extract was broken down into segments which were to be coded (65 text segments in total). The segments were provided in a table under each extract; the first column contained the segment, the second was for the answer (schema, action or world) and the third column was for comments in case the participant wanted to explain the reason for the code they assigned to any of the segments.

4.4.2.3 Procedure

The experimenter gave a presentation to the participants, in which the nature of the coding task was explained. Six practice questions were included in which participants were instructed to complete the answers for each example and answers were given to the group. Throughout the presentation, participants had the opportunity to ask questions. At the end of the presentation, participants were invited to complete the consent forms, answer the demographic questions and complete the coding task which took approximately 30 minutes. Participants were told that there was no right or wrong answer, nor was there a certain number of each code or a pattern to the coding scheme. Participants were given the option to withdraw from the study at any time. The study was held at the same time over 2 days (Tuesday and Thursday of the same week) to increase opportunity for participation. Exactly the same procedure was conducted 4 weeks after the first study, either Tuesday or Thursday. To control for extraneous variables, participants were discouraged from switching group days. All participants returned to their appropriate second study time.

4.4.2.4 Results

In the first coding session, inter-rater reliability between the participants and the criterion coder averaged 88% (range between 72% and 94%, SD = 5.5). In the second coding session, inter-rater reliability was 84% (range between 65% and 95%, SD = 10.6). Intra-rater reliability was calculated as an average of 83% (range between 60% and 92%, SD = 10).

In a review of methods measuring inter-rater reliability, Lombard et al. (2002) argued that the biggest limitation of calculating percentage agreement is the failure for it to account for agreement that would occur by chance. To account for this criticism, a random allocation procedure was employed. A random number generator was used to select the order of the 65 text segments. The three codes were assigned a number (schema = 1, action = 2, world = 3), and the number generator was instructed to generate 65 integers between 1 and 3. These numbers were assigned, in order they appeared, to the 65 segments. This procedure therefore assigned the codes generated at random, to the text segments ordered at random. This is akin to chance allocation, and the procedure was repeated five times. To assess for chance, percentage agreement between the criterion coder and the random allocation was calculated. This averaged 31% (range between 22% and 39%, SD = 6.4).

4.4.2.5 Preliminary Discussion

In the inter-rater results, percentage agreement was 88% at the first coding session and 84% in the second session. Both results are above the suggested 80% threshold for acceptable agreement. This demonstrates that different people using the PCM coding scheme on different occasions are coding CDM data in a consistent manner. Additionally, the intra-rater percentage agreement of 83% demonstrated that the same people on different occasions are also using the coding scheme in a consistent manner. These results suggest that a high level of reliability is associated with the PCM coding scheme for coding CDM data. Furthermore, the use of the random number generator to calculate chance agreement (31%) established that the results obtained from the inter-rater reliability study were considerably higher than by chance alone, increasing confidence in the use of the coding scheme. The high reliability results are not surprising when considered in the context of the theory-driven coding scheme that was implemented. Boyatzis (1998) has argued that a theory-driven code is likely to achieve consistency of judgement as interpretations are a direct commentary on the theory rather than individual interpretation. Additionally, Klein et al. (1989) suggested that CDM interviews were easier to code and led to reliable data because of their semi-structured nature and use of standardised probes.

4.5 GENERAL DISCUSSION

4.5.1 Summary

The aim of this chapter was to demonstrate how the PCM can be used as a means to explore the decision-making process by using it to structure and understand data elicited from the CDM. To achieve this, a critical incident interview was conducted with a helicopter pilot using the CDM procedure (Klein et al., 1989). As discussed in the introduction to this chapter, ADM is a complex form of NDM because there is not always a clear coupling between the decision process and the decision outcome. Therefore, it is important to understand the adequacy of the decision *process* rather than just the outcome. To do this, it is necessary to understand why actions and assessments made sense to an operator at the time the decisions were made (Dekker, 2006). Neisser's (1976) PCM offers this process understanding by placing the operator in the environment in which they work. In this study, it was demonstrated that each phase of the incident could be understood in terms of the perceptual cycle the pilot was engaged in.

In order to address the concern levelled at the method about reliability, the retest reliability of the CDM was examined. This reliability study took place over a 25-month period. It was demonstrated that the CDM was able to produce consistent interview data at T1 (9 months after the incident) and T2 (34 months after the incident). To our knowledge, this is the only study that has examined the reliability of the CDM over such a long period of time. Previous studies, for example, Taynor et al. (1987), examined reliability only up to 5 months after the incident. The results presented in this chapter inspire confidence in the use of the CDM to elicit consistent data and help to alleviate some of the concerns about the use of the method in relation to the fallibility of human memory.

It is acknowledged by Klein et al. (1989) that the qualitative data generated by the CDM require some form of subjective interpretation usually through thematic analysis of the data. Deductive thematic analysis was conducted here as themes in the data were generated from existing theory (the PCM). There is consensus in the literature that assessing inter- and intra-rater reliability is an established method to analyse the quality of a coding scheme. In this study, both inter- and intra-rater reliability were demonstrated to be above the threshold level for acceptable agreement. This suggests that the coding scheme based on the principles of the PCM is a reliable way to analyse CDM data from the perspective of understanding data in the context of Schema Theory and the perceptual cycle process of decision-making.

4.5.2 Applicability

A central tenet of the PCM is the use of schemata in information processing and decision-making. As introduced at the start of this chapter, schemata are generally advantageous, as they help organise the mass of world information available to decision makers and can reduce cognitive expenditure by directing attention and influencing action. However, they can also leave decision makers vulnerable to making schema-driven errors. Therefore, it is important to understand their role in decision-making. This is especially pertinent in the context of aviation where rigorous training procedures and SOPs are designed to structure the decision-making process by offering a limited set of choices (Simpson, 2001). It could be hypothesised that pilots will not display a high use of schemata because of the proceduralised nature of aviation. However, in this study the pilot was found to utilise a number of schemata in order to assist his decision-making process. It is possible to infer the consequences of decision-making that is largely schema driven. Generally, the use of schemata aids perception and decision-making. For example, Morris and Leung (2006) found that mental workload was not significantly increased, when task demand increased, if pilots could revert to pre-existing schemata. However, the reverse of this is that schemata can influence the production of inappropriate actions and decisions; for example, over-reliance on pre-existing but inappropriate schemata has been shown to lead to fixation on certain cues at the expense of others (Stanton et al., 2010a; Plant and Stanton, 2012a; Chapter 3). In the incident presented here, the pilot entered the wrong waypoint as a result of schema-driven decision-making. Furthermore, the majority of the diagnostic attempts were influenced by the schemata held for the aircraft's electronic system. Fortunately, in this instance, there were no adverse effects and the incident was relatively minor. However, a situation can be envisaged whereby schemata will influence decision-making at the expense of making an appropriate decision.

The main impetus for this chapter was to explore a method that could assist with understanding the role of the PCM in decision-making. The CDM was utilised as the method of qualitative data collection. The combination of the CDM for data collection and the PCM coding scheme for data analysis is, as far as the authors are aware, a novel one. Schemata are notoriously difficult to elicit (Smith and Hancock, 1995; Walker et al., 2011). Schema Theory is intuitively appealing but has not always received unanimous acclaim, which is in part due to the difficulties associated with eliciting data that can be interpreted in terms of Schema Theory (Plant and Stanton, 2013a; Chapter 2). This study has demonstrated that the probes in the CDM produce data that can be understood in terms of Schema Theory. The PCM accounts for the processes involved (i.e. why decisions made sense to the operator at the time), and it encompasses the whole system but remains human-centred by placing the operator into the environment in which they work. Literature suggests that events preceding accidents are usually non-routine or novel situations (Reason, 1990); therefore, understanding decision-making when dealing with critical incidents has the potential to shed light into the cognitive processes involved when accidents happen.

With any form of methodological exploration come lessons learnt and recommendations. Here, this primarily concerns the reliability of both the CDM and the PCM coding scheme. The test–retest reliability of the CDM was conducted to address the criticism about the reliability of the method and the fact that, to our knowledge, a retest reliability study of the CDM has not been conducted for at least 20 years and not with a time elapse of longer than 5 months (Taynor et al., 1987). Conducting a CDM retest is time consuming and may not always be practically possible if time with the interviewee is limited to one session. The study by Taynor et al. (1987) and the retest study conducted here both found the CDM to be reliable for eliciting similar data on different occasions. Furthermore, Klein et al. (1989) have argued that the use of standardised probes is likely to result in more reliable data. As such, we do not suggest that every CDM interview requires a retest; however, based on the data differences found from the retest, we recommend that time is allowed at the start of the CDM interview for practice and familiarisation with the cognitive probes. Other qualitative data collection techniques, such as verbal protocols, advocate practice as part of the procedure (Walker, 2005). Some of the differences found between T1 and T2 have been attributed to the fact that the interviewee was more familiar with the format and procedure at T2 (see Table 4.5). Therefore, a practice session to familiarise with the probes in the context of an everyday activity before the main incident interview is encouraged. Furthermore, if resources are available to conduct a retest study, it is advocated that a different researcher conducts this interview because some of the differences found between the T1 and T2 data were attributed to familiarity with the researcher (Table 4.5). The coding scheme used in this study was shown to uphold both inter- and intra-rater reliability and allowed CDM data to be analysed in relation to the perceptual cycle process. A similar coding process was employed by Plant and Stanton (2012a; Chapter 3) in relation to accident report data; therefore, it is argued that the coding scheme is mature enough to be utilised by other researchers with any form of qualitative data.

4.5.3 Avenues of Future Work

This study has presented many avenues for future work, some of which have already been alluded to throughout the chapter. One of the primary areas to focus future research concerns the level of analysis. As previously discussed, the CDM interview came from the perspective of the PNF and as such the roles of the additional crew members were not explicitly addressed. The pilot was instructed to recall the incident from the perspective of his personal decision-making process: detailed information about the decision-making of other crew members would have been highly speculative, particularly in relation to the use of schemata. Speculating on the role of other people's prior knowledge and previously held experiences in decision-making is not appropriate in the context of Schema Theory: schemata are entirely unique to the individual that holds them. Neisser (1976, p. 187) explained:

> Schemata are developed by experience; everyone's experiences are different...since every person's perceptual history is unique we should all have unique cognitive structures.

It is acknowledged that exploring decision-making solely from an individual perspective, particularly in a complex system such as aviation, is too simplistic (Plant and Stanton, 2014b; Chapter 8). However, as the motivation behind this chapter was methodological exploration, the use of one critical incident, from the perspective of one crew member, was deemed appropriate. Subsequent chapters will examine the perceptual cycle process from a larger level of analysis (see Chapters 8 and 9).

4.6 CONCLUSIONS

The value of this chapter lies in the application of an analysis method based on the PCM (coupled with the CDM as the method of data collection) to understand the processes involved in ADM. The PCM analysis structures qualitative data in such a way that the integrating elements of the PCM – schemata, actions and world information – are accounted for and attributed to decisions. It is widely acknowledged that schemata influence decision-making (Simpson, 2001), although it is less widely understood how this process happens and the potential implications of this. The use of the PCM to understand decision-making data allows for the adequacy of the decision, and decision process, to be assessed, as opposed to just the decision outcome. This aligns with the aims of NDM research (Orasanu and Martin, 1998).

In the incident presented here, potential inadequacies in the decision-making process were apparent in the pilot's reliance on his schema for an electrical fault. However, the overall decision that was made (albeit that this decision was made collectively amongst the crew), to abort the training and return to base, was entirely adequate for the situation as a successful outcome was achieved. The decision-making process, structured via the PCM, demonstrated what information (both external and internal) the pilot used to deal with the critical incident.

From this study, it would appear that CDM data can be classified within the context of the PCM and that this classification can be used to help understand and explain ADM. The motivation for such research is to develop more process-driven decision-making research in order to determine why pilots make decisions which will increase understanding about the potential consequences of those decisions. However, it is not clear how valid the PCM is as an explanation of information processing; there is no evidence in the literature of any study investigating the validity of the model itself. Therefore, the following chapter will explore the validity of the PCM as an explanation of behaviour.

5 Examining the Validity of Neisser's Perceptual Cycle Model with Accounts from Critical Decision-Making in the Cockpit

5.1 INTRODUCTION

As described in previous chapters, Neisser's (1976) perceptual cycle model (PCM) structures the interaction between a person's internal schemata (or mental templates) and the environment in which they work. Acknowledging the interaction between person and world has resulted in it becoming a popular theoretical perspective in Ergonomics research, with applications in the areas of naturalistic decision-making (NDM), situation awareness (SA) and human error to name but a few (Plant and Stanton, 2013a; Chapter 2). The model has been applied in a number of domains, including aviation (e.g. Plant and Stanton, 2012a, 2013b), road (e.g. Salmon et al., 2014a) and rail (e.g. Stanton and Walker, 2011; Salmon et al., 2013a), and it underpins Stanton and colleagues' theory of distributed situation awareness (DSA; Stanton et al., 2006; Salmon et al., 2008a, 2009a). However, the PCM has not been formally validated. Whilst the model makes intuitive sense, it is not clear whether information flows in the manner suggested by the model. This chapter utilises interview data from 20 rotary wing pilots in order to validate the PCM. In doing so, it builds on the previous chapter in which the PCM was used as the basis for examining the process of aeronautical decision-making.

5.1.1 THE PCM AND ITS APPLICATION IN ERGONOMICS

The PCM is presented in Figure 3.1; as previously described, this is an information processing model based on the idea of a reciprocal, cyclical, relationship between an operator and their environment. This ecological approach suggests information processing is cyclical rather than linear, and active rather than passive. The PCM provides a model of individual cognition based on schemata which are personal mental representations, but it recognises that cognition is extended beyond the individual because perceptions, decisions and actions are grounded within the context of the environment in which they occurred. The aviation industry has previously been criticised for studying judgement and decision-making out of the context in which it occurred, assuming that errors are the result of either technological or human failure, rather than the joint human–technology system (Maurino, 2000; Hobbs and Williamson, 2002). The recognition of both the individual and the environment has resulted in the model's popularity and endurance in Ergonomics research. The PCM explains the way in which the world constrains behaviour as well as how cognition constrains our perception of the world. The model encompasses both top-down and bottom-up information processing and explains how everyday behaviour is formed through a mixture of both approaches. As such, the model has been readily applied in the Ergonomics discipline. For example, Smith and Hancock (1995, p. 141) used the PCM as the foundation for their perspective of SA, arguing:

> …the informed, directed sampling and/or anticipation capture the essence of behaviour characteristic to SA.

Stanton et al. (2001, 2009a,b) argued that the interactive nature of the PCM is good at explaining the dynamic aspects of SA and the collective behaviour of systems as a whole, as opposed to individuals. As such, Stanton and colleagues used the PCM as one of their underpinning concepts in their theory of DSA (Stanton et al., 2006; Salmon et al., 2009a). This approach argues that awareness is distributed across human and technological agents involved in collaborative activity (Salmon et al., 2008a).

Aside from being the foundation of theoretical perspectives, the PCM has also been applied as a basis for accident analysis across a variety of domains. The schema element of the model accounts for how we act in the world and consequently can explain what can go wrong, whilst grounding this explanation in the context of the work environment. As exemplified in Chapter 3, faulty schema or faulty activation of schema can lead to erroneous performance (Norman, 1981). In the railway domain, Stanton and Walker (2011) used the PCM to explain human information processing in train driving. In the instance of the Ladbroke Grove rail crash, Stanton and Walker (2011) demonstrated that, amongst other factors, the train driver displayed a number of schema-driven errors, including mode errors (erroneously classifying the situation) and data-driven activation errors (external events that cause the activation of schemata). Similarly, Salmon et al. (2013) used the PCM to explain the actions of the truck driver who failed to comply with a level crossing, causing the death of eleven train passengers in Kerang, Australia. It was shown that the driver's failure to respond to the activated level crossing was caused by the activation of an inappropriate schema, that is a schema for the level crossing in a non-active state. This was developed through the driver's extensive experience of crossing the track when the level crossing was not activated. This then shaped the driver's perceptual exploration and he continued to cross the track as he perceived that the crossing was in a non-active state. Further environmental factors including the design of the crossing, sun glare and trees obscuring the train meant that the inappropriate schema was not overridden. Aside from averting performance errors, Neissen et al. (1999) argued that understanding schemata developed by operators can inform the design of training programmes and the programming of automation because schemata are based on operators' knowledge and skills developed through practice and are antecedents to successful performance.

5.1.2 Theoretical Validity

In its linking of the mind to action to world and back to the mind, the PCM offers a comprehensive framework, making it a popular and enduring theoretical perspective in Ergonomics. However, to our knowledge, the PCM has never been formally validated; it is assumed that the cycle of information processing occurs in the way that Neisser (1976) described it. Stanton (2002) questioned the amount of academic theory or verifiable evidence that many Ergonomics methods and models are based on. Similarly, Salmon et al. (2009b) suggested that reliability and validity should be a critical consideration when selecting methods. Methods can be reliable (provide consistent measurement) without being valid (provide an accurate reflection of a phenomenon), but they cannot be valid without being reliable (Stanton and Young, 1998, 1999). Whilst the discussion around reliability and validity is generally concerned with method selection, the importance of these in relation to theories should not be overlooked. Generally, methods and theories can either be analytical or evaluative (Stanton et al., 2013). The PCM can be considered an analytic model as it offers an understanding of the mechanisms underlying the interactions between humans and the machines they use or the environments they work in, whereas evaluative methods and theories estimate parameters of selected interactions between human and machines.

Specifically, within the context of aeronautical decision-making, O'Hare (1992) conducted an extensive review of the literature and concluded that studies on pilot decision-making have emphasised methodology rather than theory. For Ergonomists to continue to use the PCM with confidence as a theoretical foundation or an accident analysis framework, they need to be certain that the PCM offers a representative model of information processing. This is termed construct validity, which Annett (2002) defined as how acceptable a theory is and, additionally, is the test of validity that needs to be established for analytic theories (Stanton et al., 2013). However, Annett (2002) has argued that the evidence of validity is hard to come by. Likewise, Robson (2002, p. 102) stated:

> There is no easy, single, way of determining construct validity.

Assessing validity of the PCM is particularly relevant considering the use of the PCM as a framework for data derived from accident reports when first-hand accounts may not be available (e.g. Plant and Stanton, 2012a; Chapter 3; Salmon et al., 2013). In these cases, Ergonomics theories can be used to propose valid explanations of behaviour, and therefore, it is pertinent that the theory is as valid as possible to ensure the accuracy of the description of behaviour (Salmon et al., 2013).

5.1.3 Aeronautical Decision-Making: A Case Study for the Validity of the PCM

The research presented in this chapter has relevance to the validity of the PCM in general, but it is grounded in the context of aeronautical decision-making. As discussed in previous chapters, the majority of aviation crashes are attributable to decisional rather than perceptual or action errors (O'Hare, 1992). The aviation domain epitomises the NDM environment in complex sociotechnical systems, in which decisions are made in environments characterised by limited time, goal conflicts and dynamic conditions (Klein et al., 1989). In such an environment, Plant and Stanton (2013b; Chapter 4) advocated the importance of understanding the decision-making process, rather than decision outcome, in order to understand why actions and assessments made sense to the decision maker at the time they were made, that is what conditions were present at the time of the event that may have influenced performance. This approach detaches from allocating blame at the operational end of events because human performance is considered inseparable from the context in which it takes place (Maurino, 2000). Back in the 1950s, Simon argued that human rationality is bounded by design and operational features of environmental contexts (Simon, 1957, 1959). This idea has evolved into the principle of local rationality; that is, knowledge is intrinsically local and subjective so assessments and actions make sense given the operator's goals, current knowledge and focus of attention (Reason, 1990; Dekker, 2011). As such, behaviours can be explained by the situations in which they occurred. The interaction that is represented in the PCM between operator and environment allows for local rationality to be explained and understood.

This chapter seeks to answer the question as to whether people actually process information as proposed by the PCM by validating the model with decision-making data. In doing so, it also builds on the case study presented in Chapter 4 and assesses the utility of the PCM at explaining aeronautical decision-making using data from 20 pilots. Whilst this research is grounded in the context of aeronautical decision-making, it holds relevance to other complex human–machine, safety-critical, systems which characterise much of Ergonomics research.

5.2 METHOD

5.2.1 Critical Decision Method

The critical decision method (CDM; Klein et al., 1989) has been described in detail in Chapter 4. The method elicits knowledge for decisions that are devised through the recognition of critical information from the environment based on past experience and expertise. This focus on the interaction between individuals and their environment aligns the method with the ecological approach of the PCM. Furthermore, Salmon et al. (2013) suggested that the CDM is a suitable method for gathering information on schemata, which are notoriously difficult to measure (Plant and Stanton, 2013a; Chapter 2). The CDM elicits expert knowledge by asking people to discuss previous incidents they were involved with using the four-stage process described in Chapter 4. The CDM can be criticised for its reliance on verbal reports, especially as retrospective recall is required and may occur sometime after the original incident (Klein et al., 1989). To address this issue, Chapter 4 assessed the retest reliability of the CDM with a time interval of over 2 years, and it was found that there was no difference between data at time 1 and time 2 in 9 of 17 themes identified and differences in the remaining 7 themes were marginal. The consistency of information over such a long time elapse can be attributed to the use of structured CDM probes as they result in the same questions being asked, thus similar responses are elicited (Klein et al., 1989). As such, the CDM may be used with confidence that it is likely to be a reliable method for collecting data on critical decision-making.

5.2.2 Participants

Twenty rotary wing pilots participated in the research and were initially recruited through an advert placed on the British Helicopter Associated website and then via word of mouth. Males made up 95% of the sample; this is in line with data from the Civil Aviation Authority (2010), indicating that female pilots comprise 5% of all UK airline pilots, a figure that is potentially less amongst rotary wing pilots. Twenty-five percent of the sample were aged between 31 and 40 years, 40% were aged between 41 and 50 years and 35% were aged between 51 and 60 years. The pilots were all relatively experienced; flying hours ranged from 1,150 to 13,000 (mean = 5,942, SD = 3,304, median = 5,000). The pilots were employed in a range of roles, including military, personal passenger transport, North Sea transport, search and rescue and test pilots.

5.2.3 Procedure

Each pilot was interviewed at their place of work. In line with common practice using the CDM, each participant was asked to think of a critical incident they had been involved with, which was defined as being 'a non-routine or un-expected event that was highly challenging and involved a high workload in which you were the primary decision maker' (Klein and Armstrong, 2005). Most of the participants were on duty at the time they were interviewed; therefore, interview time could not be guaranteed and so the shortened version of the CDM was used; that is, only the timeline construction and deepening probe phases were conducted (see Chapter 4). Each participant provided a high-level overview of the incident and structured a timeline of events. After the incident description/timeline construction phase, the cognitive probes were asked in relation to the decision-making made during the incident. This study was granted ethical permission by the University of Southampton Research Ethics Committee. The interviews were audio recorded and later transcribed.

5.2.4 Data Analysis

The 20 CDM interviews produced data about critical incidents that were classified into five broad types of incident (see Table 5.1). The annual review by the European Aviation Safety Agency (2012) demonstrated that after loss of control incidents,

TABLE 5.1
Classification of Incident Types That Emerged from the CDM Interviews

Incident Type	Description	No. of Interviews
Technological failure or malfunction	The incident occurred as a result of a failing or malfunctioning technical system	10
Environmental conditions	The incident occurred due to poor weather conditions	4
Spurious warning/ caution	The incident occurred as a result of the onset of a warning or caution light and alarm without any other signs of a problem	3
Human error	The incident occurred as a result of human error either on the part of the pilot or someone else	2
Operational incident	The incident occurred whilst on an operational exercise and was a direct result of an operational procedure	1

systems failure (electrical and non-electrical combined) was the highest cause of fatal and non-fatal accidents in rotary wing aviation. This suggests that the breakdown of incident type demonstrated here is in line with official statistics, although, without a comprehensive comparison of incident records, it would be inappropriate to claim that the types are fully representative. The data from each interview were structured into six generic phases of incident that have been previously identified in similar data (Plant and Stanton, 2012b). The six phases were

1. Pre-incident
2. Onset of problem
3. Immediate actions
4. Decision-making
5. Subsequent actions
6. Incident containment

The data were analysed in accordance with guidelines on qualitative data analysis whereby the text was chunked into meaningful segments of approximately one sentence or less in length (Strauss and Corbin, 1990). Across the 20 interviews, there were 584 text segments. These text segments were then subjected to deductive thematic analysis which involved classifying the data into meaningful themes generated from existing theory (Boyatzis, 1998). The coding scheme was based on the three categories of the PCM and developed in line with Boyatzis' (1998) five criteria of how to structure a meaningful code (see Table 4.2). The text segments were coded for instances of the themes identified in the coding scheme. This method has been previously applied before for accident reports (Plant and Stanton, 2012a; Chapter 3) and decision-making data (Plant and Stanton, 2013b; Chapter 4).

For each interview, data from each phase of the incident were collated into a frequency table that captured 'from/to' links between the three categories of schema,

TABLE 5.2
Amalgamated from/to Link Frequency Count Data from the 20 CDM Interviews

			To				
Phase			**World**	**Schema**	**Action**	**Total Text Segments**	**% of Data/ Phase**
Pre-incident	From	World	–	5	21		
		Schema	2	–	2		
		Action	21	2	–	53	9
Onset of problem	From	World	–	20	37		
		Schema	13	–	13		
		Action	37	11	–	131	23
Immediate actions	From	World	–	15	31		
		Schema	6	–	39		
		Action	37	13	–	141	24
Decision-making	From	World	–	14	30		
		Schema	7	–	34		
		Action	38	14	–	137	24
Subsequent actions	From	World	–	10	26		
		Schema	9	–	9		
		Action	27	9	–	90	15
Incident containment	From	World	–	2	11		
		Schema	2	–	6		
		Action	11	0	–	32	5
Total	From	**World**	–	**66**	**156**		
		Schema	**39**	–	**103**		
		Action	**171**	**49**	–	**584**	**100**

action and world and was recorded in a frequency count matrix table. For example, if a text segment coded as 'schema' was followed by a text segment coded as 'action', this was recorded as one in the schema–action cell in the matrix table. This was summed across the 20 interviews to create an amalgamated frequency count for each of the six phases and across the data set as a whole. The raw data are presented in Table 5.2, and these were converted into percentages and used to populate the PCMs in Figures 5.1 through 5.7.

5.2.5 Reliability of the Coding Scheme

The reliability of the PCM coding scheme was assessed to establish the inter-rater (different people coding the data) and intra-rater (same people at different times) reliability. This is an established method to analyse the quality of a coding scheme (Burla et al., 2008) and is discussed in more detail in Chapter 4. To assess reliability, three additional coders (coders 2–4) were judged by the standard set by the expert coder (coder 1) in a blind condition; that is, raters were unaware of the expert's coding decisions. The additional coders were trained on the theory behind the coding scheme and the classification categories; they were unfamiliar with the previous use of this coding scheme. The coders were presented with 200 text segments (10 from each interview) which represented 34% of the data. The text segments were selected using a random number generator. This randomly generated 10 numbers within the range of total number of text segments for each interview. Intra-rater reliability was assessed by the three coders coding the same data 3 weeks later. Additionally, the original expert coder re-coded the 200 selected text segments. This occurred 13 months after the original coding had taken place.

Reliability scores were calculated based on percentage agreement, that is the number of agreements divided by the number of times the coding was possible, multiplied by 100 (see Chapter 4 for more detail). All results from the reliability assessment exceeded the 80% threshold level of agreement. Inter-rater reliability averaged 84% and intra-rater reliability averaged 90% (three coders) and 97% (expert coder). These results suggest that the coding scheme will generate consistent results, whether it is used by different people or the same person on different occasions and can therefore be used with increased confidence.

5.3 COMPOSITE ACCOUNT OF AERONAUTICAL DECISION-MAKING

5.3.1 Whole Incident

Figure 5.1 presents the PCM annotated with the number of from/to links (and percentages) from the amalgamated data of all of the incidents, over all of the phases. There are two possible ways that information can flow around the cycle: the traditional perceptual cycle (represented by the solid arrows around the outside of the

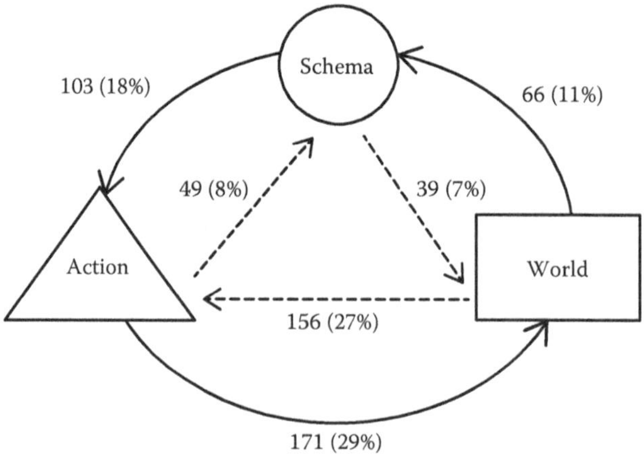

FIGURE 5.1 PCM annotated with the number of from/to links (and percentages) for the amalgamated data (n = 20).

model) and the counter-cycle (represented by the dashed arrows in the middle of the model). The traditional PCM accounts for 58% of the data, broken down as follows: 11% world–schema, 18% schema–action and 29% action–world. The counter-cycle accounts for 42% of the data, decomposed into 27% world–action, 8% action–schema and 7% schema–world.

5.3.2 Incident Phases

Composite models of each of the incident phases were also compiled. Table 5.2 displays the percentage breakdown of data across the six phases. The spread of data across the phases are as expected. The majority of the data are represented in the following three phases: onset of problem (23%), immediate actions (24%) and decision-making (24%). This is unsurprising given the nature of the interview in asking pilots to describe a critical incident and the associated actions and decisions. This is followed by subsequent actions (15%), pre-incident (9%) and incident containment (5%) phases. The pre-incident and incident containment phases provide context about what happened before and after the incident and are not directly related to dealing with the incident. Therefore, it is unsurprising that they contain less data; some pilots started their descriptions in the onset of problem phase or concluded it in the subsequent action phase, hence the smaller amounts of data in these phases.

5.3.2.1 Phase 1: Pre-Incident

The pre-incident phase contained 9% of the data. In this phase, pilots generally set the scene and described any antecedents to the incident, including the weather conditions (e.g. *it was a reasonable day, strong breeze, not a gale just wind* [CDM_018]), the operational context (e.g. *I was by myself going to pick up a client in Manchester* [CDM_008]) and any actions undertaken (e.g. *I completed the pre-take off checks* [CDM_018]). Figure 5.2 presents the PCM annotated with the number of from/to links (and percentages) from the amalgamated data of the pre-incident phase. Slightly more data flows around the traditional perceptual cycle (52%), compared to the counter-cycle (48%). The majority of data (80%) falls into a mini-cycle between world and action and back to work. This is unsurprising in the pre-incident phase where pilots described the context behind the incident and the initial piloting actions; therefore, limited data are connected to the schema node; pilots were not drawing on past experiences or expectations as they were providing a descriptive, factual, account of the conditions. The counter-cycle is discussed in more detail in Section 5.4.2.

5.3.2.2 Phase 2: Onset of Problem

Data in this phase described the critical incident (e.g. *the instrument panel lights failed* [CDM_017], *both the hydraulics systems started fluctuating...the whole aircraft started vibrating* [CDM_023], *we basically flew into a waterfall, torrential downpour...we were in inadvertent IMC in terms of loss of visual references* [CDM_011]). Figure 5.3 presents the PCM annotated with the number of from/to links (and percentages) from the amalgamated data for the onset of problem phase.

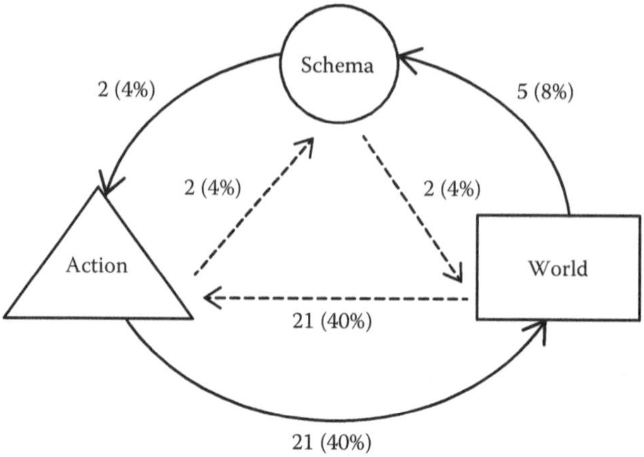

FIGURE 5.2 PCM annotated with the number of from/to links (and percentages) for the 'pre-incident' phase.

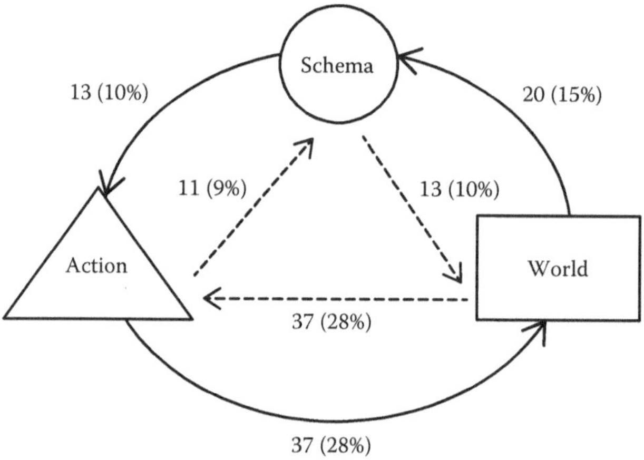

FIGURE 5.3 PCM annotated with the number of from/to links (and percentages) for the 'onset of problem' phase.

The traditional perceptual cycle flow of information accounts for 53% of the data and the counter-cycle for 47% of the data. Again, the reciprocal mini-cycle between world–action–world–action is evident, suggesting a heavier emphasis on bottom-up compared to top-down information processing. This is unsurprising given that pilots described the conditions of the incident (e.g. *the amber caution light came on* [CDM_008]) and relevant actions (e.g. *I checked the temperature and pressure gauges* [CDM_008]) in this phase. The schema, top-down, element is represented in this phase as pilots drew on past experience, either their direct or trained experience, to process what was going on (e.g. *initial perception because of what we're trained is this was a double engine failure* [CDM_009]).

5.3.2.3 Phase 3: Immediate Actions

The onset of problem phase is followed by the immediate action phase and accounts for 24% of the overall data. Whilst actions are described throughout every phase of the incident, the immediate action phase incorporates text segments that specifically relate to actions (and associated world information and schema descriptions) to ensure the immediate safety of the flight. This is exemplified through an extract from CDM_020 (including the associated PCM codes):

> …the first thing to remember is to continue flying the aircraft [action]…
>
> …we were in a turn [world]…
>
> …so I get it back to a safe flying configuration, well within the bounds of one good engine [action]…
>
> …there are minimum speeds and power figures [world]…
>
> …which you just know through training [schema]

The immediate action phase also encompasses data that described the pilot trying to identify what the problem was; this is exemplified through an extract from CDM_014:

> …immediately scanned my instruments' [action]…
>
> …saw that everything was working, readings were all normal [world]…
>
> …interestingly enough I had a similar problem about a year earlier so I had a seed in my mind about what had happened [schema]….
>
> …I shut down the engine in preparation for an auto-rotation [action]

Figure 5.4 presents the PCM annotated with the number of from/to links (and percentages) from the amalgamated data for the immediate action phase. In the immediate

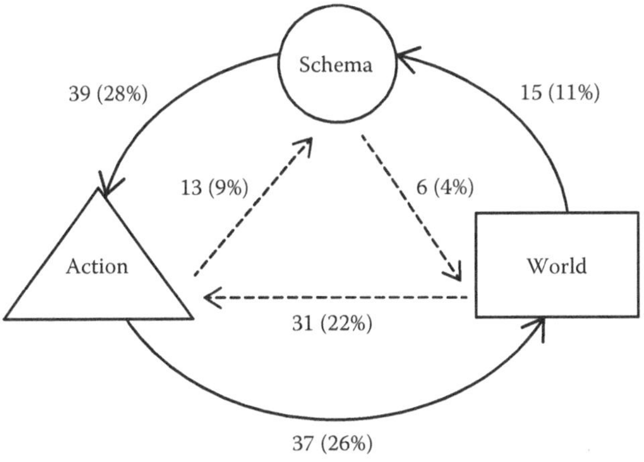

FIGURE 5.4 PCM annotated with the number of from/to links (and percentages) for the 'immediate action' phase.

action phase, the majority of the data falls into the traditional perceptual cycle (65%) compared to 35% in the counter-cycle. Furthermore, compared to other phases this phase is very top-down driven with 28% of data in the schema–action link. This phase is characterised by an iterative cycle of assimilating information in the world and drawing on past experiences, whether that is from training or direct experience, to produce the appropriate actions. The excerpt from CDM_014 presented earlier exemplifies this transition around the traditional PCM (action–world–schema–action).

5.3.2.4 Phase 4: Decision-Making

The decision-making phase accounts for 24% of the total data. Decisions are made continuously throughout the incident, but this phase is characterised by text segments specifically relating to the overall decision made for dealing with the incident. Data for this phase generally came from the answer to the CDM probe: *what was the primary decision that you made?* Text segments in this category relate to the actions taken during this phase, the use of past experience and the environmental information that assisted the decision-making process. Therefore, all three elements of the PCM were represented in this phase, with the majority of the data falling into the traditional perceptual cycle (63%) compared to the counter-cycle (37%). Figure 5.5 presents the PCM annotated with the number of from/to links (and percentages) from the amalgamated data for the decision-making phase. As with previous phases, within the counter-cycle the world–action link is strong, contributing to a reciprocal mini-cycle between world–action–world–action. The schema–action link accounts for a quarter of the data in this phase, suggesting a strong top-down information processing component in the decision-making phase. Many of the pilots drew on past experiences to assist their decision-making process; this ranged from direct past experience, e.g. *I'd had a similar incident...there was a bang, nowhere near as loud as this...but a little bit of vibration associated with it, which suggested to me*

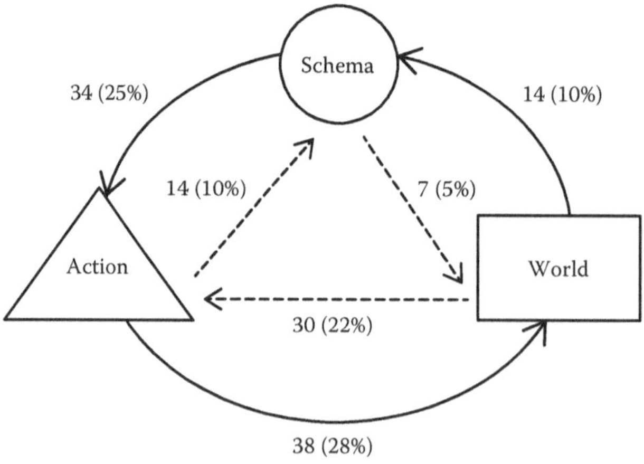

FIGURE 5.5 PCM annotated with the number of from/to links (and percentages) for the 'decision-making' phase.

it could be the same problem (CDM_014), past experience developed through training, e.g. *the decision of what to do was in my experience because of training, I had seen this instance before in a simulator* (CDM_004) and vicarious past experience developed through reading or speaking about potential situations, e.g. *something did go through my mind, I had been talking to an instructor who asked me what I would do in a similar situation...* (CDM_006).

5.3.2.5 Subsequent Actions

The subsequent action phase accounts for 15% of the overall data. This phase encompassed data relating to additional actions taken as a result of the decision-making phase. Generally actions involved whether to land the aircraft immediately, land the aircraft as soon as possible or resolve the incident in the air. Figure 5.6 presents the PCM annotated with number of from/to links (and percentages) from the amalgamated data for the subsequent action phase. There is almost an even split between the two cycles: traditional perceptual cycle (51%) and the counter-cycle (49%). Within both cycles, the action–world (30%) and world–action (29%) links account for most of the data, suggesting that this phase is characterised by bottom-up information processing. This makes sense in a phase that focuses on the actions undertaken and the world information that is drawn upon to confirm the actions. For example:

> ...I was scanning for a field in front [action]...
>
> ...the field had to be somewhere in my field of view, not the best or ideal field but the best field directly in front of me in a mile range [world]...
>
> ...I pointed my aircraft roughly at a set of fields...I was trying to reduce the speed [action]...
>
> ...the dials were moving all over the place [world] (CDM_005)

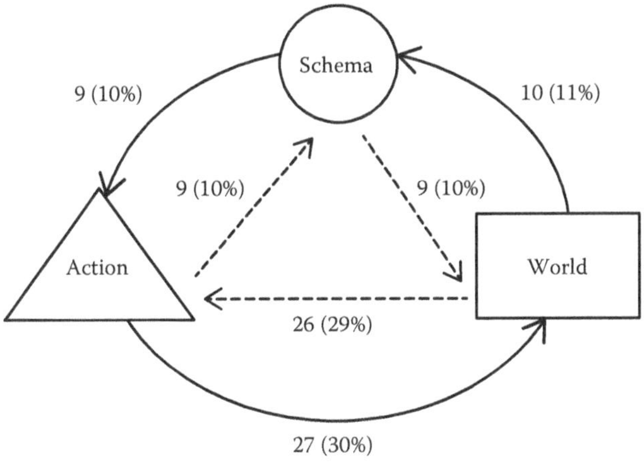

FIGURE 5.6 PCM annotated with the number of from/to links (and percentages) for the 'subsequent actions' phase.

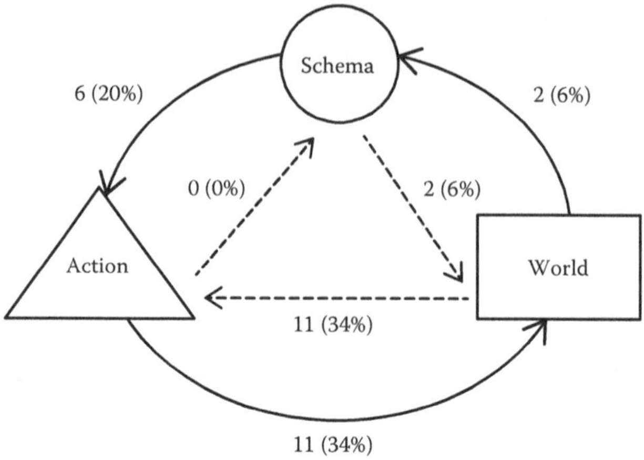

FIGURE 5.7 PCM annotated with the number of from/to links (and percentages) for the 'incident containment' phase.

5.3.2.6 Incident Containment

The final phase of the incident is the incident containment phase. This phase only accounts for 5% of the overall data because it was a short phase, generally containing only a few text segments as pilots concluded their account of the incident. This phase is exemplified with the following text segments: *it was a normal shut down because the transmission had not yet seized* (CDM_008), *landed and another aircraft landed next to me, engineer looked at it...* (CDM_014), *I was talked in by a controller...I broke cloud at 200 feet and landed normally* (CDM_006). Within this phase, the traditional perceptual cycle represents the majority of the data (60%) compared to the counter-cycle (40%), which is incomplete as the action–schema link is not represented (see Figure 5.7).

5.4 DISCUSSION

The aim of this chapter was to explore the validity of the PCM in order to establish whether the data flows in the manner described by Neisser (1976). To undertake this analysis, 20 CDM interviews were conducted with rotary wing pilots about critical incidents they had been involved with. Each incident was structured into six key phases, and the text segments within each phase were subjected to deductive thematic analysis. The data were amalgamated to produce composite PCMs of information flow for each of the incident phases and across the data set as a whole. In exploring the validity of the PCM, this research also built upon the previous study in Chapter 4 in which the PCM was used as the basis for examining the process of aeronautical decision-making. Within this section, the practical applications are discussed, along with an evaluation of the methodology and future research endeavours are outlined.

5.4.1 Validating the PCM

The principle of validity refers to whether a concept or theory corresponds to the phenomenon observed in the real world. In this case, whether the cycle of information processing in the perceptual cycle occurs as Neisser (1976) described it. The PCM suggests that information flows in a cyclical manner in which the environment (world information) informs the operator, modifying their knowledge (schema). These schemata direct the operator's activity (action) in the environment. These actions result in the sampling of the environment, which in turn informs the operator. Here, the study sought to determine the construct validity of the PCM by conducting deductive thematic analysis on qualitative data obtained from 20 CDM interviews. Thematic analysis is the process of identifying, analysing and reporting patterns within data (Braun and Clarke, 2006). This was deductive, or theoretically driven, because a classification scheme based on the three key elements of the PCM was used to analyse the data (schema, action and world). Braun and Clarke (2006) advocated the use of deductive thematic analysis when coding for specific research questions, in this case exploring the validity of the PCM. Braun and Clarke (2006) argued that the hallmarks of a quality thematic analysis include limited overlap between themes, all data are able to be classified and all aspects of a theme should adhere to a central idea or concept. Here, the classification scheme resulted in exhaustive (i.e. all text segments were classified) and exclusive (i.e. there was no ambiguity about which code text segments should be assigned to) coding, achieving Braun and Clarke's (2006) principles of quality thematic analysis. This suggests that the PCM is effective at explaining critical decision-making and therefore supports the previous assertions made in Chapter 4 that the PCM is a useful framework from which to study the processes of (aeronautical) decision-making.

A frequency count of the from/to links was conducted on the text segments, as they appeared in the interview transcripts, for example, schema–action or world–action. This was used to create composite perceptual cycles of each of the phases of the incident (see Figures 5.2 through 5.7). The composite analysis demonstrates that, on the whole, the PCM explains the critical decision-making data. The traditional perceptual cycle in which information flows from world to schema to action and back to world accounted for 58% of the data. This demonstrates that the majority of data flows around the perceptual cycle in the way Neisser (1976) anticipated; however, there is a noteworthy proportion of the data that flow in the counter-cycle (42%), particularly in the world–action link. This results in a mini-cycle between world to action and then action back to world; this is discussed in detail in Section 5.4.2. Overall, the links that come from/to the schema node holds less data than the links relating to the action and world nodes. This is potentially a product of the data collection method or the experience level of the participants and is accounted for in Section 5.4.4.

The traditional perceptual cycle flow of information is most evident in the immediate action (65%) and decision-making (63%) phases of the incident. The immediate action phase is characterised by an iterative cycle of assimilating information in the world and drawing on schemata, developed through training or direct experience, to produce appropriate actions. Similarly, in the decision-making phase text segments were related to actions taken, the use of past experience and the available environmental information that assisted the decision-making process. The analysis of 20 case studies has demonstrated that the perceptual cycle theory can explain the CDM data, exclusively and exhaustively.

5.4.2 Counter-Cycle and Levels of Behaviour

As described in the previous section, the traditional perceptual cycle was able to account for the majority of the critical incident data. However, it was also found that some of the data fitted into the counter-cycle, with information flowing in the opposite direction from the traditional world–schema–action cycle and instead flowing in a world–action–schema cycle. This counter-cycle accounted for 42% of the overall data; it was most evident in the subsequent action phase, accounting for 49% of the data. Within the counter-cycle, the majority of the data occurred in the world–action branch. In the traditional PCM, the world is only connected to action by schemata. An equal amount of data was often found in both the world–action branch and the action–world branch. Initially, when analysing the data from each incident a flow analysis was conducted in which the from/to links were recorded around the PCM in the order in which they occurred. The main point taken from this flow analysis is that a reciprocal mini-cycle often occurred between the world and action nodes; this is exemplified later with data from CDM_019:

> We launched from the pick-up site… [action]
>
> The cloud was still lowering [world]
>
> …given the high risk I thought I would head to the top of the cloud, so we climbed 4500 ft and zoomed over on clear sky [action]
>
> Now we were over the destination, there was no gap in the cloud [world]
>
> We were going around in circles [action]
>
> I had maps but there was no radar [world]
>
> I was trying to calculate how much clearance I had [action]

Rasmussen's (1983) skill-, rule- and knowledge-based behaviour (KBB) taxonomy (SRK) can help explain this. SRK describes three different levels of cognitive control that can be exercised by an individual over their actions. Skill-based behaviour (SBB) refers to the smooth execution of highly practised, largely physical actions where there is virtually no conscious monitoring. Klein et al. (1989) described this automaticity as 'recognition primed'. At this level, recognition of certain patterns (world information) maps directly onto the selection of an appropriate response (action). SBB develops through extensive practice in which operators develop cued response patterns suited for specific situations. In the from/to link analysis, SBB manifests as direct links from world information to actions (as exemplified earlier). For skilled pilots, in familiar aircraft and environments, the action of flying the aircraft is executed largely without effortful thought, hence the schema node is bypassed as information from the environment is automatically assimilated and appropriate actions are performed (Rasmussen, 1974). This can also be considered in terms of the supervisory attentional system (SAS) proposed by Norman and Shallice (1986). They proposed that for well-learned action sequences, two levels of control are possible: deliberate conscious control and automatic processing. The SAS relates features from the environmental situation to a set of schemata. As schemata become active, responses are selected and performed. Practice of the response results in a closer coupling of feature-to-schema-to-response, producing automatic behaviour (see Chapter 2).

Neisser's (1976) original PCM distinguished between top-down and bottom-up information processing – the former being schema-driven processing around the left-hand side of the PCM and the latter being environmentally driven information processing depicted on the right-hand side of the model. To generate appropriate responses to novel situations, an operator needs to integrate more general knowledge concerning the behaviour of the system, the characteristics of the environment and the goals to be achieved (Drivalou and Marmaras, 2009). In the PCM, this manifests as the top-down flow of information processing, whereby actions are driven by stored schemata and the resulting change on world information is monitored. This KBB requires considerable feedback and this is accounted for in the PCM by the modifying effect that world information can have on the future utilisation of schemata. In the data analysis, the schema node is the least represented with less from/to links when compared with the action and world nodes. The participants in this study were experts in the sense that, based on flying hours, they were experienced aviators. By its definition, a schema is a mental template developed through past experiences and associated expectations (Bartlett, 1932). As such, experts have more past experiences and therefore a wider repertoire of schema to draw upon. It could therefore be hypothesised that the schema links should be highly represented in the data because the expert pilots utilise their stored schema to deal with the incidents. However, when viewed from the perspective of the SRK taxonomy, it is more likely that the expert pilots engage in automatic SBB rather than the effortful KBB. Furthermore, automatic routines (SBB) may continue to control behaviour, even when higher-level processing becomes necessary so that a person engaged in RBB or KBB could still be engaged in actions under skill-based control (Hobbs and Williamson, 2002).

Hobbs and Williamson (2002) have argued that the SRK taxonomy has begun to acquire status as an 'industry standard' in a variety of settings. However, the SRK taxonomy only focuses on the mind of the individual that is, the level of cognitive control that is implemented, and it does not acknowledge the role of the environment in shaping and modifying operators' behaviour. The fact that the PCM acknowledges the interaction between an operator and their environment is a significant strength of the model. Furthermore, it aligns the PCM with the cognitive systems engineering (CSE) approach (Hollnagel and Woods, 1983; Hollnagel, 2001) which is a relatively recent branch of the Ergonomics discipline. CSE views the unit of analysis as the whole system, that is the human–machine interaction, rather than just the decomposition of both the human and the system as separate entities, something clearly echoed in the PCM (see Chapter 2 for a more detailed discussion of the role of the PCM in CSE). Plant and Stanton (2014a; Chapter 7) have previously justified the use of the PCM to explore decision-making processes over the more popular recognition-primed decision (RPD) model (Klein et al., 1989) because the RPD model does not capture the cyclical interaction between operator and their environment as successfully as the PCM does. As such, the explanatory power and potential utility of the PCM (and in turn the SRK taxonomy or the RPD model) can increase if they can be aligned together as each capture elements that the others do not.

5.4.3 Practical Applications

At the start of this chapter, it was argued that it is essential to establish the validity of models and theories so that they can be used with confidence (Stanton and Young, 1998; Salmon et al., 2009b). This is especially pertinent with the PCM as it has been extensively applied and used for accident analyses where first-hand accounts are often unavailable (e.g. Plant and Stanton, 2012a; Chapter 3; Salmon et al., 2013). The results of this study have demonstrated that the PCM does uphold the principles of construct validity and as such it can continue to be used with increased confidence. Interestingly, the findings also demonstrated the occurrence of a counter-cycle within the original model; this needs to be accounted for if the PCM is used as a framework for data analysis. The results presented here have offered an extension to Neisser's original work, and whilst the research context is aeronautical decision-making, it can be argued that aviation epitomises the NDM environment in complex sociotechnical systems. The findings have the potential to generalise to other complex human–machine systems, either in other transport domains, control process domains and other safety-critical systems.

O'Hare et al. (1992, 2010) argued that decision-making performance is not a function of factual knowledge and that alone; a repertoire of perceptual motor skills and formal rules is insufficient to become a fully proficient pilot. Previous experience is a benchmark of expertise, regardless of domain, experts possess the ability to compare current situations with previously experienced situations. The role of schemata is captured by the PCM, and from the excerpts provided in Section 5.3, it is clear that pilots utilised a variety of schemata, including those developed through direct past experience, trained past experience and vicarious past experience. The PCM therefore has the potential to be a framework for training aids centred on decision-making. Within the aviation domain, the importance of case-based reflection is well established as a means of enhancing aeronautical decision-making. For example, Henley et al. (1999) advocated the use of reflective journals as effective means of accelerating the development of an experiential knowledge base from which to make decisions and judgements. Similarly, O'Hare et al. (2010) found that participants who reflected on a set of cases involving pilots flying into adverse weather conditions were more likely to follow weather-related flight rules in a simulated flight than participants who completed a free recall task. The PCM framework could potentially be utilised as a way to structure reflective decision aids or decision-making training modules. This research has demonstrated that pilots engage in the PCM when dealing with critical incidents. Therefore, if pilots are taught the principles of the PCM and the interaction between the elements, there is the potential for higher retention rates if reflection and/or training is aligned with the information processing mechanism pilots are actually utilising.

5.4.4 Evaluation of Methodology and Future Research Endeavours

Some of the methodological considerations with research of this nature have already been acknowledged in previous chapters, and they will be further discussed in this section. In relation to the data collection method, the CDM is open to criticism. Of the three elements of the PCM, schemata are the hardest to elicit and measure. They are mental constructs and are therefore not open to direct observable measurement and are not necessarily as consciously recalled as world information might be (Plant and Stanton, 2013a; Chapter 2). The justification for selecting the CDM was well considered and provided in Chapter 4; however, there is still a potential for bias in the data (e.g. text segments to relate to world and action rather than schema). The results have been previously discussed in light of the SRK behaviour taxonomy, and it might be that schemata were not elicited by the CDM because of the expertise of the participant group. An alternative explanation is that the CDM method is not sophisticated enough to elicit schema-based data and the results were partly a product of the data collection method. This can only be resolved through comparing and contrasting different data collection techniques. Annett (2002) argued that data collection is always open to some form of bias. Furthermore, specifically in relation to mental models, Revell and Stanton (2012) showed that bias varies considerably depending on the theoretical perspective used and methods of data capture and argued that as long as biases were acknowledged then findings could be trusted.

In relation to the data analysis method, Braun and Clarke (2006) have warned that because deductive thematic analysis is driven by the researcher's theoretical interests, it is more explicitly analyst driven and potentially more subjective than inductive analysis in which themes emerge from the data set. As such, a reliability assessment was undertaken to establish the inter- and intra-rater reliability of the coding scheme. The results demonstrated average inter-rater reliability of 84% agreement and intra-rater reliability of 90% agreement, increasing confidence in the use of the coding scheme. Here, all possible measures were taken to optimise the objectivity of the analysis: the coding scheme was developed in line with Boyatzis' (1998) five criteria for how to structure a meaningful code, a systematic analysis procedure was implemented that followed the guidelines provided by Braun and Clarke (2006) and an assessment of reliability was conducted.

In their overview of the advantages of the deductive thematic analysis method, Braun and Clarke (2006) argued that the approach can usefully summarise key features of a large body of data and can generate unanticipated insights. This was clearly evident in this study as the data analysis uncovered the counter-cycle and the mini-cycle occurring between elements of the PCM there were previously unlinked. This has led to questions about the nature of engagement in the perceptual cycle between novices and experts, and lines of further enquiry are clearly evident.

5.5 CONCLUSION

This chapter sought to explore the validity of the PCM and, in doing so, built on previous research supporting the use of the PCM as a framework for understanding, modelling and explaining decision-making processes (as described in Chapters 3 and 4). Based on a thorough review of the literature, it is assumed that this is the first attempt to establish the validity of a widely used model and the motivation for such research is to ensure the model can continue to be used with confidence, although more studies of this nature need to be undertaken in a wide variety of domains. A model or theory can be considered valid if, after careful scrutiny, no objection or contradiction can be sustained. This research has demonstrated that on the whole the PCM provides an accurate representation of information processing when dealing with critical incidents. However, the counter-cycle that was observed suggests that experts might process information in a slightly different way, and this warrants future investigation and provides a new development for the theory. Throughout this chapter, different categories within each of the PCM elements have been alluded to. For example, in Section 5.3, reference was made to different types of world information (e.g. natural environment, operational context and location), different types of schemata (e.g. direct past experience, vicarious past experience and trained past experience) and different types of actions (e.g. aviating, navigating and communicating). Currently, the PCM categorisation provides a high-level overview of what is going on, but it would be useful to explore the detail in each category to establish the types involved and explore differences, be that in terms of incidents, pilot experience or domains of application. Chapter 6 aims to further develop the PCM theory by considering the high-level elements of schema, action and world in more detail.

6 Development of a Perceptual Cycle Classification Scheme

6.1 INTRODUCTION

The motivation for this chapter was to build on the previous work and continue to develop the utility of Neisser's (1976) perceptual cycle model (PCM) as an explanatory tool for decision-making behaviour. There are two aims of the research. The first aim is to provide detailed descriptions of the three categories of the PCM in order to deliver a more complete understanding of the decision-making process. This is achieved in the development of the schema–action–world (SAW) taxonomy using data from 20 decision-making interviews. The second aim is to use this taxonomy to provide a detailed explanation of aeronautical critical decision-making (ACDM).

6.1.1 Role and Utility of Taxonomies in Ergonomics Research

Taxonomy is the science of defining groups, traditionally biological organisms, on the basis of shared characteristics and giving names to those groups. The primary purpose of classification is to describe relationships of the constitute elements in regard to each other and to similar elements, and to simplify these relationships so that general statements can be made about classes of elements (Fleishman and Quaintance, 1984). Developing and utilising taxonomies to assist research efforts can have a number of advantages, including the facilitation of efficient and accurate data collection and analysis, establishing common terminologies that can be employed by a variety of researchers, and they are associated with practical benefits such as training design and performance measurement (Fleishman and Quaintance, 1984; Moir et al., 2003). Taxonomies can be used in two main ways; some researchers choose to develop a bespoke taxonomy for the domain in question, and examples of this include a gliding accident analysis taxonomy developed by Jarvis and Harris (2010) and a construction work taxonomy developed by Moir et al. (2003). Alternatively, a pre-existing taxonomy can be applied to relevant data. This approach was taken by Stanton and Salmon (2009) who applied well-established taxonomies of human error to develop an understanding of automotive driver errors.

In Ergonomics research, taxonomies can be used for a variety of purposes including classifying behaviour of individuals or teams, predicting risk with error taxonomies and for accident analysis purposes. As acknowledged by Salmon et al. (2012), accidents and accident causation are central themes in global Ergonomics research efforts. The methods used to investigate and analyse accidents are critical in aiding understanding of underlying causes and indicating areas for systemic improvements. Taxonomies have been widely applied for this purpose, particularly in aviation (for comprehensive reviews, see O'Hare (2000) and Jarvis and Harris (2010)) which led Griffin et al. (2010) to affirm that taxonomic accident models are greater in number than any other. In this chapter, a bespoke, data-driven, perceptual cycle taxonomy was developed and is presented in the context of decision-making. Fleishman and Quaintance (1984) argued that a comprehensive taxonomy should focus on human and environmental dimensions, rather than just one or the other, as this allows the impact of each on the other to be understood. The SAW taxonomy developed here fulfils this requirement as it captures the three elements of the PCM: internal cognitive schema, actions undertaken and world information available.

6.1.2 PCM as a Process-Oriented Approach to Aeronautical Decision-Making

Aviation epitomises the naturalistic decision-making (NDM) paradigm (Klein et al., 1989) as it is an environment in which experts make rapid decisions with available information, in time-critical situations and that can have negative payoffs. Decision judgement errors are often a significant contributory factor in incidents and accidents, with some arguing that they are the primary factor in over 50% of general aviation accidents (Simpson, 2001). For example, O'Hare et al. (1994) found that 63% of fatal and serious aviation accidents in their data set were characterised by decision errors. Decision-making research presents an interesting field of enquiry because there is often no clear standard of correctness, and there can be a loose coupling between event outcome and decision process (Orasanu and Martin, 1998). The decision maker acts according to their understanding of the situation at the time, a principle that has been termed local rationality (Reason, 1990; Dekker, 2011). Errors can therefore arise because of deficiencies in the decision makers' knowledge base or in the process of reaching a decision. As such, it is essential to understand why the actions and assessments undertaken by a decision maker made sense to them at the time (Dekker, 2006).

The PCM emphasises the role that both schemata (internally held mental templates) and world information play on shaping actions, including decisions. Schemata are triggered by contextual conditions and direct perception and behaviour, thus interaction in the world. The world information that is experienced can have a modifying and updating effect on cognitive schemata, thus influencing further interaction. This individual model of information processing recognises that cognition is extended beyond the individual because behaviours are grounded within the context of the environment in which they occurred. As described in previous chapters, it offers a useful framework for explaining the mechanisms involved in decision-making. The model has been most widely applied as an explanatory framework to understand factors such as situational awareness or decision-making from the perspective of individuals operating as part of larger systems. Whilst the domains of application and methodological approaches vary, there are two distinct commonalities. Firstly, the research has usually occurred in situations where data are derived from accident reports in which first-hand accounts of incidents are not available (Salmon et al., 2013). Therefore, it is essential that the explanation of decision-making (or any element of behaviour) provided by the PCM is as accurate and detailed as possible because it is used to propose valid explanations of behaviour. Secondly, the research efforts have attempted to understand what schemata were held by the operators, what environmental information they were exposed to and how these factors interacted to produce their actions and behaviour within the context of their operating environment. However, to our mind, one of the biggest limitations of the PCM is that in its current form it offers only a very high level of descriptive detail, comprising only the three categories of schema, action and world. Overcoming this issue is one of the key aims of this research and will be achieved through the development of an extended PCM taxonomy.

6.2 METHOD: TAXONOMY DEVELOPMENT

6.2.1 Data Collection

The CDM was used to collect data from helicopter pilots about decision-making during critical incidents. This method has been described in detail in the preceding chapters. The data in this chapter utilised the same data set from Chapter 5. For a detailed description of the participants and procedure, refer to Sections 5.2.2 and 5.2.3.

6.2.2 Data Analysis

6.2.2.1 Data Treatment

Section 5.2.4 describes the data treatment procedure. In accordance with the guidelines on qualitative data analysis, the text was chunked into meaningful segments of approximately one sentence or less in length. In Chapter 5, this resulted in 584 text segments: if a text segment coded as 'world' was followed by a text segment coded as 'world', this was combined into one count for 'world' (rather than two). However, in this chapter, because the interest is in the categories within the high-level codes, each individual occurrence of the high-level code was counted, so in the previous example this would have resulted in two counts for 'world'. This resulted in 904 text segments across the 20 interviews. The data analysis techniques undertaken here are based on the principles of thematic analysis and have been described in detail in Chapter 4. The data were subjected to both deductive and inductive thematic analysis.

6.2.2.2 Deductive Thematic Analysis

Braun and Clarke (2006) argued that there is not one ideal theoretical framework or method for conducting qualitative analysis, but what is important is that the framework or method matches the researcher's focus of investigation. The first stage of the analysis process was deductive thematic analysis using the existing PCM theory (see Chapter 4 for a detailed description). The three coding themes were

1. Schema (statements relating to the use of prior knowledge and knowing things because of experience or expectations)
2. Action (statements of doing an action or discussing potential actions that could be taken)
3. World (statements relating to potential or actually available information in the world including physical things, conditions or states of being)

The focus of this chapter is not to look at these themes in any detail because this has been done previously (see Chapters 4 and 5) but rather to explore themes within these high-level categories via inductive thematic analysis.

6.2.2.3 Inductive Thematic Analysis

Inductive thematic analysis was undertaken on the data in each of the three high-level categories of schema, action and world in order to uncover more detailed themes within these data. Inductive thematic analysis is the process by which the data are used to generate themes (Patton, 1990). In its purest form, inductive analysis is a process of coding data without trying to fit it into a pre-existing coding frame (Braun and Clarke, 2006). However, the data in this study were already classified into the three PCM codes, and this therefore had some bearing on the nature of the themes generated in the inductive analysis process. An advantage of thematic analysis is the flexibility of the method which allows themes to be determined in a number of ways. Braun and Clarke (2006) stressed the importance of being consistent with the analysis when generating themes. Here, the constant comparison technique was employed whereby each text segment was compared with previous items to see whether the same or a different phenomenon was described. Furthermore, the text segments were exclusively coded, meaning that only one theme could be applied to one text segment. If a text segment represented more than one theme, the segment was split up accordingly (e.g. if two different types of action were described in one segment), but this did not affect the high-level coding. The taxonomy was developed through an iterative process of review and refinement using the opinions and expertise of colleagues in the research group. The process of inductive analysis resulted in the identification of six schema subtypes, 11 action subtypes and 11 world subtypes. These are presented in the SAW taxonomy in Section 6.3.1.

6.2.2.4 Relationship and Frequency Analysis

Aside from providing a more detailed explanation of the PCM, this research sought to explore the relationships between different elements of the perceptual cycle. To do this, for each CDM interview, each code was collated into a frequency table that captured 'from/to' links between the different categories as they appeared in the coded transcripts. For example, a text segment coded as 'action_decision action' (from), followed by a segment coded as 'world_standard operating procedure' (to), was recorded in the frequency matrix. This was summed across the 20 interviews to create an amalgamated frequency count for each of the 6 phases and across the data set as a whole. This frequency count analysis was subjected to network analysis using the Agna™ software. This is a social network analysis tool but is becoming an increasingly popular method for general network analysis. Walker et al. (2010) argued that it is particularly compatible with the distributed cognition perspective because it focuses on relationships amongst operators embedded in their work context. It provides a range of different metrics for analysing networks, and these are listed and described in Table 9.1.

Specifically, the metric of sociometric status was of most interest and was used to define key information elements related to critical decision-making. Sociometric status refers to the relative importance of a node (concept) within a network as its calculation is based on the connectedness (i.e. number of connections to other nodes) of a particular information element. The argument is that concepts with high sociometric status values represent key concepts as they are highly connected to other concepts within the network (Salmon et al., 2014b; Stanton, 2014). Here, the concepts (i.e. PCM subcategories) with a sociometric status value above the mean plus one standard deviation for the network were identified as primary concepts, those with a value higher than the mean but lower than the mean plus one standard deviation were identified as secondary concepts and those with a value lower than the mean were identified as tertiary concepts. The network analysis also enabled a PCM network of critical decision-making to be produced. Section 6.3.2 describes the results of this analysis.

6.2.3 Reliability Assessment

As conducted in previous chapters, the reliability of both coding schemes was assessed by calculating inter- and intra-rater reliability. It is well acknowledged that calculating reliability is an essential component for evaluating the quality of a coding scheme or taxonomy (Fleishman and Quaintance, 1984; Burla et al., 2008). A detailed description of reliability and justification for calculating percentage agreement is provided in Chapter 4. In this study, to assess reliability, three additional coders were judged by the standard set by the expert coder in a blind condition; that is, raters were unaware of the expert's coding decisions. The high-level coding scheme has been previously evaluated in Chapter 5. It was found that all results from the reliability assessment exceeded the 80% threshold level of agreement. Furthermore, Chapter 4 demonstrated that agreement of 20 coders exceeded the 80% threshold, and this was shown to be considerably higher than from chance alone.

To assess the reliability of the SAW taxonomy, the coders were given a 30-minute session describing the different classification categories. The coders were presented with 200 text segments (10 from each interview) which represented 22% of the data. The text segments were selected using a random number generator. This randomly generated 10 numbers within the range of total number of text segments for each interview. Additionally, the original expert coder re-coded the 200 selected text segments. This occurred 13 months after the original coding had taken place. In the initial analysis, the schema and world subcategories met the 80% threshold level of agreement for both inter- and intra-rater reliability (schema inter-rater, 87%;

intra-rater, 88%; and world inter-rater, 82; intra-rater, 86%). However, the action subcategory only reached the threshold for intra-rater reliability (82%), the inter-rater agreement was 73% and does not meet the threshold level of agreement.

On close examination of the data, it appeared that the subcategory in which there was most disagreement was 'mental action'. In the original coding scheme, this was defined as *statements relating to actions that are not explicitly observable, actions that occur in the mind (e.g. thinking about something or imagining how a course of events might unfold)*. From the reliability assessment, it was clear that this category of mental action was too ambiguous and was easily confused with other categories including decision action and concurrent diagnostics. With the collaboration of the reliability coders, it was decided that 'situation assessment' would be a more appropriate classification, and this was defined as *statements relating to actions that relate to the evaluation and interpretation of available information*. All text segments relating to mental action were re-coded by the coders and were either assigned to the situation assessment category or another relevant category (e.g. concurrent diagnostics). The re-analysis of this category produced an inter-rater reliability agreement of 84%, above the accepted threshold. This process of review and refinement of themes is an essential part of thematic analysis and demonstrates the benefit of conducting a rigorous reliability study as it resulted in a higher-quality coding scheme.

6.3 RESULTS

6.3.1 SAW Taxonomy

The impetus for this research was to provide a more detailed account of the high-level categories of schema, action and world that are provided in Neisser's (1976) original PCM. This was achieved through a process of inductive thematic analysis to produce a data-driven taxonomy. The SAW taxonomy resulted in 28 categories: 6 schema subtypes, 11 action subtypes and 11 world subtypes.

6.3.1.1 Schema Categories

Table 6.1 presents the six schema subtypes along with a description of the classification theme and examples from the text extracts of the CDM interviews. The six schema subtypes include vicarious past experience, direct past experience, trained past experience, declarative schema, analogical schema and insufficient schema. Of the entire data set, 15% related to the overall category of schema; the breakdown of the schema data across the subtypes is also provided in Table 6.1. Direct past experience, trained past experience and declarative schema are the most represented categories, accounting for 26%, 25% and 24% of the data, respectively.

6.3.1.2 Action Categories

Table 6.2 presents the 11 action subtypes that were identified through the inductive analysis process. The 11 action subtypes included aviate, navigate, communicate, system management, system monitoring, environment monitoring, concurrent diagnostics, decision action, situation assessment, non-action and standard operating procedure. Of the entire data set, 43% were categorised as action and the breakdown

TABLE 6.1
Schema Subtypes

Schema Subtype	Description	Examples	Data (%)
Vicarious past experience (VPE)	Statements relating to experiencing something in the imagination through the description by another person (e.g. hearing a colleague recall an incident they were involved with) or documentation (e.g. reading about a certain event in an industry magazine or incident/accident report).	*I knew I had surged the engine…I had heard about surging…I hadn't experienced it but I knew that the engine was surging. It had been described to me, in books. You don't train for it. No one plays you a sound clip, it's more by reading documents I suppose.*	5
Direct past experience (DPE)	Statements relating to direct personal experience of similar events or situations in the past. This covers events experienced in live, operational contexts as opposed to those experienced through training.	*I have experienced levels of vibration on other aircraft and I know what is normal, what is abnormal, this exceeded it tenfold…*	26
Trained past experience (TPE)	Statements relating to knowledge developed by direct personal experience of a specific task, event or situation, experienced within the confines of a training scenario (e.g. ground school training, simulator training or training sorties).	*The decision of what to do was in my experience because of training. I had seen this instance before in a simulator.*	25
Declarative schema (DS)	Statements relating to a schema that manifests as a descriptive knowledge of facts, usually as a product of the world information available.	*I knew it had just come out of maintenance, I was aware it could be a spurious event.*	24
Analogical schema (AS)	Statements relating to comparisons between things for the purpose of explanation and clarification. Typically, these analogies will be structural analogies of physical objects or states of affairs in the world (akin to mental map or mental model).	*How high am I, how fast am I, can't see a lot so having to make this picture in my head based on the information that I do know.*	3
Insufficient schema (IS)	Statements relating to inadequate or lacking knowledge, i.e. a schema is not developed for a certain situation.	*I didn't have a mental picture of the trends and fluctuations.*	17

Note: Theme name, description, examples and percentage of data within category.

TABLE 6.2
Action Subtypes

Action Subtype	Description	Examples	Data (%)
Aviate (Av)	Statements relating to direct manipulation (handling) of flight controls in order that the aircraft can be flown and safety is maintained	*I attempted to roll the aircraft level.*	23
Navigate (Nav)	Statements relating to the process of accurately ascertaining position and planning and following a route or desired course	*I followed the coast back.*	2
Communicate (Comm)	Statements relating to the sharing or exchange of information	*I transmitted a non-standard mayday call.*	9
System management (Sys Man)	Statements relating to the processes of making an input into technological systems in order that the interaction or manipulation has an explicit output	*I put in St. Albans head into the navigation system, So I typed in the three digit code which is St. Albans head.*	8
System monitoring (Sys Mon)	Statements relating to looking at (observing, checking) displays to gain an understanding of the situation	*I did a complete scan of all the systems information.*	10
Environment monitoring (Env Mon)	Statements relating to observing or checking the internal or external physical environment in order to establish the current state of affairs	*I was keeping eyes out for ground contact and searching for visual references.*	4

(Continued)

TABLE 6.2 (*Continued*)
Action Subtypes

Action Subtype	Description	Examples	Data (%)
Concurrent diagnostic action (Conc Diag)	Statements relating to the process of determining, or attempting to determine, the cause or nature of a problem by examining the available information at the time the incident is occurring	*We initially started looking for circuit breakers, to look if any had popped.*	6
Decision action (DA)	Statements relating to a conclusion or resolution that is reached after considering the available information	*The first decision was to idle back the bad engine, rather than shut it down.*	16
Situation assessment (Sit Ass)	Statements relating to actions that relate to the evaluation and interpretation of available information	*Trying to take into account the threats to you and the aircraft, i.e. if I precede down a given path what is it likely to result in?*	8
Non-action (Non A)	Statements relating to actions that were not performed, either because the situation did not warrant a particular action or because equipment faults did not allow a particular action to be performed or because the pilot made an error or omission	*I couldn't read any of the instruments or communicate.*	5
Standard operating procedure (SOP)	Statements relating to following the prescribed procedure that ought to be routinely followed in a given situation	*I completed the pre-take off checks.*	9

Note: Theme name, description, examples and percentage of data within category.

of these data across the action categories is provided in Table 6.2. Aviate (the action of flying the aircraft) is the most represented category (23% of data), followed by decision action (16%).

6.3.1.3 World Categories

Table 6.3 presents the 11 world subtypes that were identified through the inductive analysis process. The 11 world subtypes included natural environmental conditions, technological conditions, communicated information, location, artefacts, display indications, operational context, aircraft status, severity of problem, physical cues and absent information. The world category made up 42% of the total data set; the breakdown across the individual world categories is provided in Table 6.3. The categories that shared the highest proportion of data were natural environmental conditions, location and operational context; each contained 14% of the world data.

6.3.2 Perceptual Cycle Analysis of ACDM

The motivation behind developing a detailed taxonomy based on the PCM is that it increases the explanatory power of the model in order that a more detailed understanding of how operators engage in the perceptual cycle is provided. Here, this has been explored from the perspective of ACDM. Calculating the sociometric status of each of the SAW taxonomy categories produced primary, secondary and tertiary concepts for dealing with a critical incident; this was based on data summed across the 20 CDM interviews and therefore provides a composite account. Figure 6.1 shows the PCM annotated by primary, secondary and tertiary concept for the whole process of dealing with a critical incident, that is, not phase of flight specific. There were five primary concepts identified overall – in the action category, these were aviate and decision action, and in the world category, these were location, natural environment and display indication. There were no primary concepts identified in the schema category. Eight secondary concepts were identified – in the action category, these were system management, communicate and situation assessment; in the world category, these were operational context and physical cues; and in the schema category, these were direct past experience, trained past experience and declarative schema. The remaining 15 concepts from the SAW taxonomy were identified as tertiary concepts.

The sociometric analysis was also conducted for each of the six phases of dealing with a critical incident; the results of this are provided in Figure 6.2. A detailed explanation of the incident phases is provided in Chapter 5. For each phase, an overview based on the primary and secondary concepts is provided:

- *Phase 1: Pre-incident* – In this phase, the pilots set the scene and described the antecedents to the incident. The most relevant concepts in this phase generally came from the world concepts. There were two primary concepts, both subtypes of world, being natural environmental conditions and operational context. Furthermore, location and physical cues were also defined as secondary concepts. Aviate and standard operating procedures were highlighted as the most relevant action concepts and declarative schema was the most important schema concept.

TABLE 6.3
World Subtypes

World Subtype	Description	Examples	Data (%)
Natural environmental conditions (NEC)	Statements about natural environmental conditions (e.g. weather, light, temperature, noise).	*Fortunately it was a clear day, nice sunny day.*	14
Technological conditions (Tech Cond)	Statements relating to the state of technological artefacts (e.g. with regard to appearance and working order).	*...engines responded and all other stuff came back on.*	8
Communicated information (Comm info)	Statements relating to information available to the pilot from other people (e.g. other crew members, ATC, coastguard).	*I received the cloud base report from Newquay.*	5
Location (Loc)	Statements relating to particular places or positions.	*...so now we were over the destination.*	14
Artefacts (Art)	Statements discussing physical objects, including written information, symbols, diagrams or equipment.	*I had the flight reference cards.*	5
Display indications (Dis Ind)	Statements relating to the information elicited from the physical artefacts.	*Only thing identifiable was the high transmission oil temperature.*	12
Operational context (Op Cont)	Statements relating to the routine functions or activities of the organisation (e.g. search and rescue, police search, military training). This can include statements about the importance of being serviceable for the operational context or crew familiarity with the aircraft and how this effects decision-making.	*The aircraft was relatively heavy because we were taking people back to Germany.*	13

(*Continued*)

TABLE 6.3 (*Continued*)
World Subtypes

World Subtype	Description	Examples	Data (%)
Aircraft status (Air Stat)	Statements relating to the current status of the aircraft's integrity or performance (e.g. how good or bad it is flying, the current configuration of the aircraft, autopilot activation).	*The aircraft was flying fine.*	7
Severity of problem (Prob Sev)	Statements relating to how bad (or otherwise) the critical incident is.	*We weren't in any immediate danger.*	3
Physical cues (Phys Cue)	Statements relating to external cues that provide information of conditions or states of being (e.g. noises, sounds, vibration, smells).	*There was a loud bang, coughing and spluttering.*	11
Absent information (Abs Info)	Statements relating to information that was missing, not present or lacking. Reasons for this may include technical faults with equipment or non-existent information.	*I didn't have a comprehensive map.*	8

Note: Theme name, description, examples and percentage of data within category.

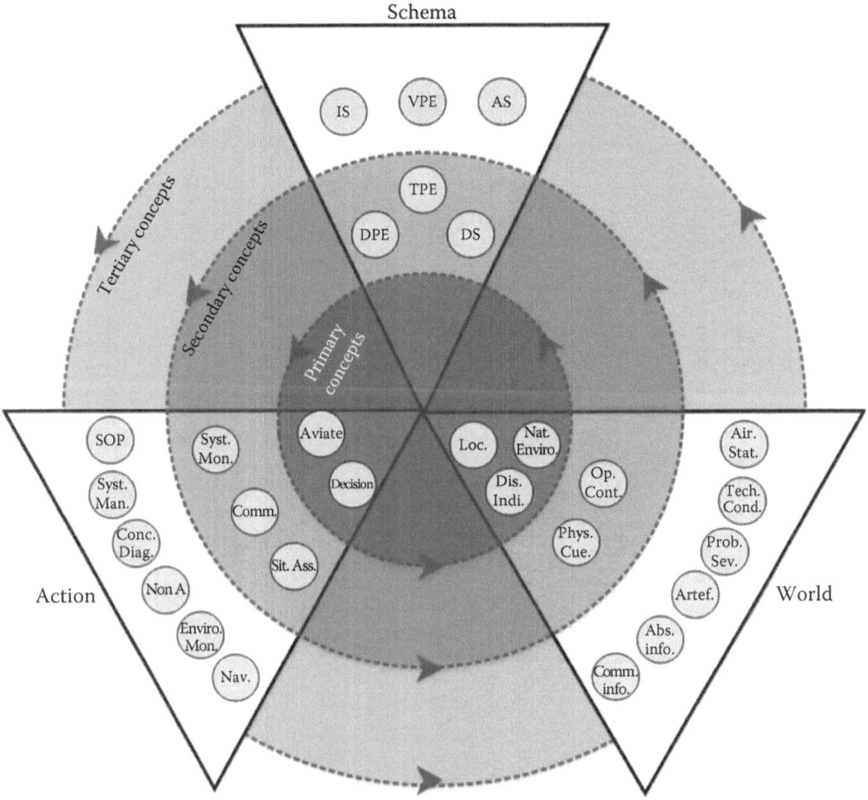

FIGURE 6.1 The primary, secondary and tertiary concepts of the PCM for dealing with a critical incident.

- *Phase 2*: *Onset of problem* – This phase was characterised by the primary concept of physical cue. Technological conditions and display indication follow as the second and third most relevant world concepts. In this phase, the most relevant action concept and third most relevant overall concept was aviate, the act of flying the aircraft. Other important action concepts included systems monitoring, concurrent diagnostics and systems management. The most relevant schema concepts were direct past experience, trained past experience and insufficient schema.
- *Phase 3*: *Immediate actions* – There were no primary concepts in this phase, but the most important concept was display indication (world concept) followed by trained past experience (schema concept). Action subtypes generally dominated this phase, with 7 of the 14 secondary concepts coming from the action category, including aviation, systems monitoring, concurrent diagnostics, decision action, communicate, situation assessment and standard operating procedures.

PCM	SAW Taxonomy	Phase of Incident: Pre-Incident	Onset of Problem	Immediate Actions	Decision-Making	Subsequent Actions	Incident Containment
Schema	Vicarious past experience						
	Direct past experience		5	5	4	8	
	Trained past experience		6	2	7	11	
	Declarative schema	6			9	3	
	Analogical schema						
	Insufficient schema		11	11		9	
Action	Aviate	4	3	3	10	1	1
	Navigate						
	Communicate			8	6	4	5
	System management		12				4
	System monitoring		8	4			
	Environment monitoring					10	
	Concurrent diagnostics		9	6			
	Decision action			7	1	2	
	Situation assessment			9	3		
	Non-action			12			
	SOP	5		10	12		

FIGURE 6.2 Matrix showing the relative importance of the SAW concepts in each phase of dealing with a critical incident. *(Continued)*

		Phase of Incident					
PCM	SAW Taxonomy	Pre-Incident	Onset of Problem	Immediate Actions	Decision-Making	Subsequent Actions	Incident Containment
World	Natural environment	1	7			6	
	Technological conditions		2				
	Communicated information						
	Location	3			2	12	2
	Artefacts			13			
	Display indications		4	1	11	5	
	Operational context	2				13	3
	Aircraft status				5	7	6
	Severity of problem						
	Physical cues	7	1	14			
	Absent information		10		8	14	

Key: Primary concept ■ Secondary concept ■ Tertiary concept Not represented □

Note: Numbers denote order of importance for primary and secondary concepts.

FIGURE 6.2 (*Continued*) Matrix showing the relative importance of the SAW concepts in each phase of dealing with a critical incident.

- *Phase 4*: *Decision-making* – In this phase, unsurprisingly, decision action was the primary concept. The remaining secondary concepts were relatively evenly spread around the three elements of the PCM, with four world concepts (location, aircraft status, absent information and display indication), four action concepts (situation assessment, communication, aviate and standard operating procedure) and three schema concepts (direct past experience, trained past experience and declarative schema).
- *Phase 5*: *Subsequent actions* – There are no primary concepts in this phase, but the top five most relevant secondary concepts are aviate, decision action, declarative schema, communicate and display indication.
- *Phase 6*: *Incident containment* – In this phase, again, there are no primary concepts and only action and world subtypes appear as secondary concepts. Included in the action category were aviate, systems management and communicate and in the world category location, operational context and aircraft status. Eleven of the 28 concepts do not feature in this phase of the incident, given that only 5% of the data are represented in this phase.

6.4 DISCUSSION

Decision-making can be the deciding factor as to whether normal situations turn into incidents and incidents turn into accidents. As argued by Salmon et al. (2012), accidents represent a complex systems phenomenon because they are the result of the interaction between causal factors that reside at all levels of a system. As such, it is appropriate to view decision-making through the lens of distributed cognition rather than as a process that occurs in the head of an isolated individual. Neisser's (1976) PCM provides this distributed cognition perspective by acknowledging the interaction between internal cognitive schemata held by the decision makers and the external environment in which decisions are made. However, to date, the PCM has only provided a high-level description of the three elements that make up the model. The impetus for this chapter was to decompose these three elements into more detailed subtypes in order that the descriptive power of the model is increased. In doing so, a more comprehensive understanding of ACDM is provided.

6.4.1 SAW Taxonomy

The SAW taxonomy is a 28-item classification scheme for gaining a more detailed understanding of the three high-level categories of the PCM and therefore a comprehensive view of the decision-making process. The taxonomy can be utilised to explore the interaction of the specific concepts during the perceptual cycle process. The development of this taxonomy was data driven using interview data from 20 helicopter pilots which generated over 900 text segments. Inductive thematic analysis was employed to generate the taxonomy items which were continually reviewed and refined. The resulting SAW taxonomy consisted of 6 schema, 11 action and 11 world categories. It enables researchers to classify data according to the perceptual cycle but in a much more detailed way than previously possible. The value of this is that the taxonomy considers factors that are internal to the decision maker (i.e. cognitive schemata) but also external factors that have the potential to influence the decision-making process (i.e. environmental information) and the actions that mediate between the two. This comprehensive taxonomy can therefore be used to explore the interaction between an operator and the world in which they work, which is essential for understanding decision-making processes, rather than just the decision outcome. Understanding this process has the potential to lead to more detailed explanations of incidents and accidents rather than just a surface-level label of 'human error' or 'bad decision-making' (Dekker, 2006).

6.4.2 Gaining Perceptual Cycle Insights into ACDM

As stated by Fleishman and Quaintance (1984, p. 44):

> Classification is not an end in itself, rather it is a tool that provides an increased ability to interpret, predict or control some facet of performance.

The SAW taxonomy was developed to gain a more detailed understanding about the process of ACDM from the perspective of the PCM. The analysis discussed is a composite account summed across the 20 interviews and is considered both as a whole and by phase of incident. In both instances, the SAW concepts were scored on sociometric status to determine their relative importance (the higher the status, the more important the concept). This resulted in the classification of primary, secondary and tertiary concepts (only primary and secondary concepts are discussed as these are most pertinent to the ACDM process). O'Hare (1992) argued that little effort has been put into developing descriptive models of aeronautical decision-making. With this in mind, Figure 6.3 provides the SAW model of ACDM based on the primary and secondary concepts identified in the data. The links between each concept represent the directional flow of information and the strength of the links was informed by the Agna network data – the thicker the line, the stronger the link is between the concepts. The analysis showed that the most important concept for ACDM was the action concept of 'aviate'; highlighted in Figure 6.3 as most of the other concepts are connected to aviate in some way. This is unsurprising given that aviate is the primary task management requirement in the 'aviate–navigate–communicate–manage systems' strategy employed by all pilots when dealing with non-normal situations. It is of paramount importance that, regardless of what is happening around them, pilots continue to fly their aircraft. In a study of task management priorities, Schutte and Trujillo (1996) found that participants prioritised 'aviate' when dealing with non-normal situations. The next most important concept was 'decision action'. This is to be expected in data collected for decision-making research and increases confidence with the method for collecting decision-based data. The next three most relevant concepts were from the world category: location, natural environmental conditions and display indications.

The absence of any schema subtypes in the primary concepts suggests that ACDM is predominately driven by bottom-up information processing; that is, the environmental information that is received drives the actions that are undertaken. This is supported by the ACDM model (Figure 6.3) which shows the strongest

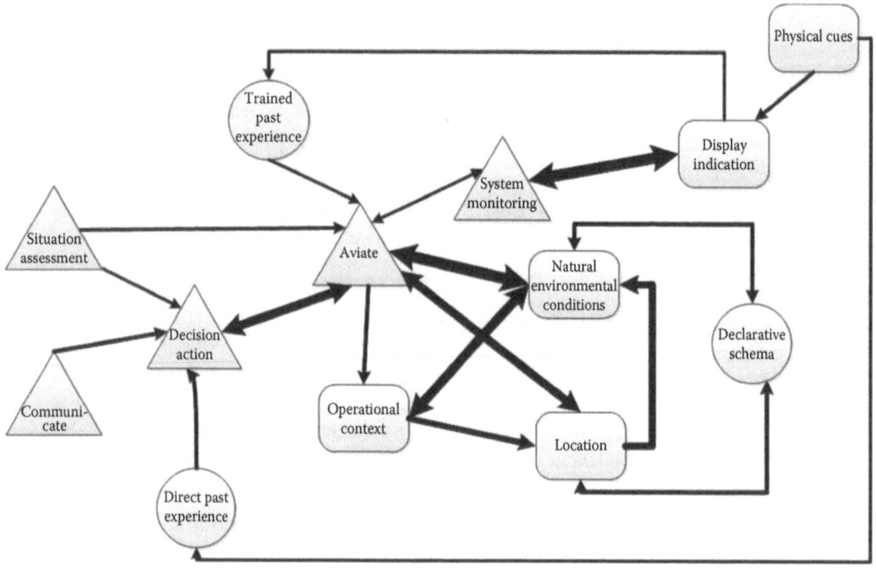

FIGURE 6.3 SAW model of aeronautical critical decision-making. *Note*: For clarity, only primary links are depicted.

links between aviate and the world concepts of location and natural environment. This information processing strategy is akin to what Rasmussen (1983) described as skill-based behaviour (SBB), that is the smooth, automatic, execution of highly practised physical actions. Chapter 5 describes this in detail, but in summary SBB is the smooth execution of highly practised, physical actions, with virtually no conscious monitoring. Rasmussen (1983) also described rule-based behaviour (RBB) and knowledge-based behaviour (KBB), the former requiring identification and recall of known rules which are stored in memory and the latter being effortful, conscious processing of unfamiliar situations. Through the process of training and learning, pilots will progress from KBB, through RBB, to SBB in the flight control task (aviating). However, the RBB and KBB levels will manifest themselves in more unfamiliar situations, such as dealing with critical incidents. Three schema subtypes appear in the secondary concepts: direct past experience, trained past experience and declarative schema. There is a strong link between direct past experience and decision action (Figure 6.3); this can be considered a manifestation of KBB as operators draw on their internally held schemata to develop solutions to unfamiliar situations.

The most dominant model in NDM research is Klein's (1998) recognition-primed decision (RPD) model. In summary, this captures how experts make decisions based on recognition of past experiences that are similar to the current situation (see Chapter 2). The RPD model does not depict the reciprocal interaction between the person and their environment to the level that the PCM does (see Plant and Stanton, 2014a; Chapter 7). In Figure 6.3, it is clear that environmental factors (in the form of natural environmental conditions, location, display indications, physical cues and operational context) play a key role in the overall ACDM process. The RPD model is not able to account for these factors in the decision-making process. Some of the schema and action concepts depicted in Figure 6.6 are accounted for in the RPD model via its depiction of 'perceived as typical' (akin to direct past experience or trained past experience) and 'action'/'implement course of action'. However, the reciprocal links between these elements and the environmental elements are missing.

The phase-specific analysis is summarised in Figure 6.2. The SAW concepts vary in importance according to phase of incident; due to space constraints, three phases will be discussed: the *onset of problem* phase is characterised by the physical cues and technological conditions that generally alerted the pilot to the problem. Aviate is the most relevant action, in line with the previous discussion about the importance of maintaining adequate handling of the aircraft at all times. The other action subtypes that feature as secondary concepts are systems monitoring, concurrent diagnostics and systems management. These are actions associated particularly with technological failure which comprised 50% of all the critical incidents in the data set. Similarly, in Schutte and Trujillo's (1996) study the majority of non-normal situations were based on systems failures, and it was found that pilots focused their attention on systems 46% of the time. The range of both world and schema subtypes in the secondary concepts suggests that pilots are engaging in both bottom-up and top-down information processing to deal with the onset of the problem, that is drawing on environmental information and their stored knowledge to troubleshoot what might be happening. In the third phase, *immediate actions*, display indication was the primary concept, and the next 11 secondary concepts were subtypes of the schema and action categories. This suggests that this phase was characterised by top-down information processing which is supported by the importance of trained past experience and direct past experience as secondary concepts. Insufficient schema also features as a secondary concept which was apparent when pilots talked about insufficient background knowledge to deal with the problem that was presenting. The most relevant actions are characteristic of a phase that involves ensuring the immediate safety of the aircraft, including aviate, systems monitoring, concurrent diagnostics and situation assessment. The immediate action phase is followed by the *decision-making* phase where, unsurprisingly, decision action is the primary concept. Secondary concepts include a variety of all three subtypes as text segments in this phase related to the actions undertaken once the primary decision was made (e.g. situation assessment, communicate and aviate) and how these actions were influenced by the world information (e.g. location, aircraft status and absent information) and stored knowledge (e.g. direct past experience, trained past experience and declarative schema). The decision-making phase of the incident is the only phase in which all PCM subtypes are represented in the data.

6.4.3 Evaluation

Fleishman and Quaintance (1984, p. 2) argued:

> In order to constitute a useful tool, a scientific classification must be founded inductively upon relationships that have a factual basis, i.e. relationships must be derived directly from the data.

The SAW taxonomy was data driven using an inductive thematic analysis process to develop it as objectively as possible. However, as Braun and Clarke (2006) acknowledged, researchers can never fully free themselves from their theoretical perspectives. So whilst the SAW taxonomy is data driven, it was still developed within the confines of the PCM, and the data had already been deductively classified based on the high-level categories of this model. Mitchell et al. (2014) conducted an extensive review of human factors taxonomies for use in a medical setting and found that a third of the taxonomies they reviewed did not have a theoretical model underpinning their development and that almost all of the taxonomies either focused on cognitive or contextual factors, rarely both. The SAW taxonomy was underpinned by a theory that considers both the contextual and cognitive factors of behaviour. Overall, the SAW taxonomy was developed with as little manipulation as possible, and as Revell and Stanton (2012) argued, as long as biases are acknowledged, then trust can be placed in the research output. Fleishman and Quaintance (1984) defined three ways of evaluating classification systems. The first is internal validity, that is, that the system is logical and parsimonious within itself. Within this, they identified reliable descriptors, mutually exclusive classes and exhaustive classifications as requirements of internal validity. The reliability assessment demonstrated that a reliable taxonomy has been produced and all of the 904 text segments were classified exclusively; that is, a text segment was placed into only one category, and exhaustively (i.e. every text segment was able to be classified).

The second way to evaluate taxonomies is by establishing external validity; that is, the taxonomy is capable of accomplishing its intended purpose. In this case, that is increasing the explanatory power of the PCM by providing a more detailed understanding of the categories within this model and their role in decision-making. The taxonomy needs to demonstrate usefulness as an aid in interpreting and integrating research results and establishing whether new data sets can be accurately assigned to the categories. Establishing external validity can only be achieved through subsequent research studies, ideally by other researchers to limit potential biases. The structured nature of taxonomies that is provided by the classification guidance for the attributes of each category (e.g. Tables 6.2 and 6.3; Figure 6.2) increases reliability when used by different researchers. Although the flip side of this is that analysts are constrained by what they can categorise (Salmon et al., 2012). Work presented in Chapter 9 starts this process of validation.

The third and final way that Fleishman and Quaintance (1984) identified to evaluate taxonomies is the principle of utilitarian criterion or use rate; that is, the taxonomy is actually used by researchers in the fields of interest. This can only be achieved through future research efforts; a potential barrier to achieving this is the aviation specific nature of the taxonomy. The SAW taxonomy is specific to understanding ACDM from the perspective of the PCM; this approach maximises specific utility of the method, although this is at the expense of generalisability to other problem domains (Fleishman and Quaintance, 1984). However, out of 28 categories only 2 categories make reference to aviation terminology (aviate and aircraft status). These could easily be supplemented by the domain under analysis as the descriptions for these categories are general enough to apply to any system interaction, particularly other transport domains. In domains other than aviation, it may also be necessary to alter the category of standard operating procedures (potentially to rules or conventions).

Previous chapters have identified the role of the perceptual cycle in ACDM, but from this work the finer detail of the interactions was able to be explored. The results make intuitive sense, insomuch that concepts have the most relevance in the phases where they would expect to be found and are less relevant in phases where they are not expected. For example, decision action is the primary concept in the decision-making phase, whereas severity of problem and concurrent diagnostics do not feature in the pre-incident phase (because the incident had not happened yet). For exploratory studies, intuitive sense is important because it points towards the appropriateness of the data collection and analysis methods. The relative importance of concepts was objectively determined by the sociometric analysis function in Agna which suggests that the development of the SAW taxonomy and general classification method employed was appropriate for gaining an increased understanding about ACDM. Furthermore, the results are similar to those obtained by Schutte and Trujillo (1996) who employed a different methodology, so it would suggest that a real phenomenon is described, thus increasing confidence in the content validity of the taxonomy. With regard to the SAW model of ACDM (see Figure 6.3), network representations have been gaining increasing prominence as a method for representing complex sociotechnical interactions (e.g. see Griffin et al., 2010; Walker et al., 2010; Plant and Stanton, 2014a; Chapter 8). Moving away from linear representations by applying networks to aviation incidents allows for a more comprehensive understanding of the context in which decision were made and the interactions that occurred (O'Hare, 2000; Griffin et al., 2010).

6.4.4 Future Applications and Research

The potential exists to utilise the SAW taxonomy in any qualitative data analysis including data collected from interviews, communication transcripts or verbal protocols. Stanton and Salmon (2009) stated that a valid taxonomy can be used either proactively to anticipate potential situations or retrospectively to classify and analyse situations after they have occurred. The SAW taxonomy could potentially be used to identify, a priori, decision-making components and to provide a detailed classification of decision-making processes in the event of incidents or accidents. Both applications have associated practical implications, for example in the identification of areas that may benefit from decision-making training or redesigning interfaces to facilitate the provision of information in the external environment.

Fleishman and Quaintance (1984) argued that a comprehensive taxonomy should enable attributes that distinguish a category member from the general population to fall into a meaningful pattern, given knowledge of relevant literature, such as contrasting expert and novice perspectives. For example, O'Hare et al. (1998) utilised the CDM in a study of white-water rafting guides and found that experts used more analogical reasoning compared to novices. With perceptual cycle processing, it is likely that experts would rely on schemata to select relevant environmental cues and therefore be more selective than novices in the information they attend to. Inexperienced and experienced pilots are known to have different accident types, the former being associated with loss of control accidents due to lack of basic handing skills and the latter being associated with more cognitively driven accidents, such as complacency (O'Hare et al., 1994; Civil Aviation Authority, 1997). Using the SAW taxonomy to highlight where and how differences in decision-making manifest between different demographic groups will increase its external validity and provide useful information about ACDM.

The SAW taxonomy also has the potential to complement existing methods. In a review of system-based accident analysis methods, Salmon et al. (2012) concluded that Accimaps, one of the most popular accident analysis methods, was limited by its lack of taxonomies at each descriptive level; for example, it recognises flawed decisions but without identifying the factors that influence them. Salmon et al. (2012) also argued that the use of taxonomies facilitates the aggregation of data across multiple cases; this is consistent with Fleishman and Quaintance's (1984) assertion that taxonomy development allows for the establishment of a base for conducting and reporting research studies to facilitate their comparison. This is supported by the fact that in their review, Salmon et al. (2012) found the STAMP method, which is taxonomy based, provided a clearer understanding of why erroneous and inappropriate decisions were made. This was assisted by the fact that STAMP considers both the context in which decisions were made and how flawed mental models contribute to decisions, akin to the schema–world distinction in the SAW taxonomy. It is not being claimed that the SAW taxonomy in its current form is an alternative accident analysis method. But the review by Salmon et al. (2012) highlighted the benefits of taxonomy-based accident analysis methods and the gaps in existing methods. From this, there appears to be the potential for the SAW taxonomy to complement existing methods, and this should certainly be an avenue for future research considerations.

6.5 CONCLUSIONS

The PCM has been extensively applied to understand decision-making processes in a variety of domains. Traditionally, however, it has only been defined at a high level of abstraction. The work presented here has developed a classification scheme to provide a more detailed description of perceptual cycle processes. In doing so, it is envisaged that this will enable a more complete understanding of decision-making to be achieved. The development of Ergonomics theory is essential if Ergonomics is to be conceived as a discipline in its own right. A challenge for the development of a unified theory is the cover-all requirement, that is, to predict a huge variety of systems interactions in an assortment of contexts (Hancock and Diaz, 2002). The PCM has potential utility as a unifying theory of Ergonomics because it is capable of providing domain-independent insights in systems interactions (Plant and Stanton, 2013b). The PCM, with its description of the reciprocal relationship that exists between the mind and the environment, bridges the gap between ecological and information processing theories of behaviour. For the PCM to realise its potential as a unifying theory, research efforts need to focus on theoretical development and associated methods. The work presented here in the development of the SAW taxonomy has begun to address this and this will be continued in Chapter 7, where a method for perceptual cycle elicitation is presented.

7 Schema World Action Research Method for Understanding Perceptual Cycle Processes

7.1 INTRODUCTION

The perceptual cycle model (PCM) has been widely applied to understanding processes such as decision-making and situation awareness as it provides a distributed approach, acknowledging the role of both internally held mental schemata and external environmental information in shaping these processes. There are, however, no associated methods with this model, and therefore, researchers have had to be creative with other methods in order to infer perceptual cycle processing. This chapter presents the development and testing of the Schema World Action Research Method (SWARM), an interview schedule for eliciting perceptual cycle information. Application of the method is exemplified through a case study of an aviation critical incident. The results demonstrate high theoretical validity with the PCM, and a test–retest reliability assessment demonstrated high reliability when the method was applied on two separate occasions. Future research endeavours are acknowledged in which the method is expanded across domains.

7.1.1 PERCEPTUAL CYCLE AS AN EXPLANATORY FRAMEWORK

As has been extensively discussed in previous chapters, the PCM places emphasis on the processes involved in decision-making, as opposed to just considering the decision output. This is important for considering local rationality; that is, a decision maker acts according to their understanding of the situation at the time, and it is essential to understand why certain actions and assessments were taken (Reason, 1990; Dekker, 2006). What in hindsight can be labelled as 'errors' often arise because of deficiencies in the decision makers' knowledge base or in the process of reaching a decision. In the highly automated aviation environment, more procedural and predictable tasks are handled by machines, whilst humans are left responsible for tasks that require diagnoses, judgement and decision-making (Militello and Hutton, 1998). This is an increasingly common trend in other safety critical industries, and as such, it is more pertinent than ever to undercover the decision-making process that humans engage in.

In recent years, Neisser's PCM has seen resurgence in popularity; owing to the fact that it places the human decision maker within the operational context, that decisions are made and therefore considers cognition-in-context (see Chapter 3). This more distributed view of decision-making aligns with contemporary Ergonomics research that has diverged from individual considerations, to focusing on the interaction of systemic factors (Salmon et al., 2015). As described, the PCM has been widely applied as an explanatory framework in a variety of domains to understand how individuals operate as part of larger systems. To date, however, there has not been a structured methodological approach for eliciting perceptual cycle information, particularly the schema element. In Chapter 2, it was acknowledged that this has traditionally been a barrier to the application and utility of Schema Theory and will be discussed further in the next section.

7.1.2 Eliciting and Representing Perceptual Cycle Processes

A fundamental component of the PCM are schemata; these are akin to internal mental templates and therefore cannot be directly measured but only inferred through the manifestation of observable behaviour or recalled information. Early studies of inferring schemata from recall and reproduction of information were valuable at defining some of the key features of schemata, but the artificial nature of studies gave little away about schematic processing in the real world. Rather, naturalistic methods that elicit qualitative data are more suited to the study of schemata in the real world. Contemporary approaches for eliciting schemata, and interpreting the perceptual cycle process, pair data collection methods (such as verbal protocols and interviews) with qualitative data analysis (such as network representation or thematic analysis). These are forms of cognitive task analysis (CTA) as they are approaches that determine the cognitive elements (i.e. mental processes and skills) required for task performance and the changes that occur as skills develop (Militello and Hutton, 1998; O'Hare et al., 1998).

In CTA, the majority of data collection usually occurs through interviews (Militello and Hutton, 1998). One of the most popular in Ergonomics research is the critical decision method (CDM; Klein et al., 1989, Klein and Armstrong, 2005). This is conducted retrospective of task performance, and therefore, it is not associated with the same disadvantages of think-aloud techniques, in terms of the potential for changing the nature of the task (Stanton et al., 2013), although it has its own limitations such as memory derogation and the questionable ability of people to verbalise cognitive behaviour. However, Militello and Hutton (1998) argued that verbal reports are no more or less problematic than any other methodology. The CDM uses semi-structured cognitive probes to understand decision-making during non-routine situations resulting in a transcript that can then be analysed. Previous research presented in this book has demonstrated that CDM transcripts can be effectively analysed to understand perceptual cycle processes by using a deductive thematic analysis technique. Specifically, the CDM focuses on eliciting knowledge for behaviours Klein et al. (1986) classed as recognition-primed decisions, that is decisions for which alternative actions are derived from recognition of critical information and prior knowledge.

The recognition primed decision (RPD) model is the most popular and enduring model in the naturalistic decision-making domain. However, it is not without its critics. For example, Lipshitz and Shaul (1997) argued that it was too simplistic and did not adequately acknowledge the role of schemata in decision-making, and Plant and Stanton (2014a) argued that it does not go far enough at acknowledging the interaction between schemata and environmental information and the modifying effect each can have on the other. This is supported by O'Hare et al. (1998), who argued that the model fails to adequately describe the situational understanding that decision makers gather from their environment. The explanation provided by the RPD model is primarily one of the decision-making processes occurring in the head of the decision maker, and consequently, the CDM is focused on eliciting this content knowledge of operators. Whilst the role of the environment is acknowledged, this is only at a very high level with questions such as 'what information cues were attended to' and 'what information was considered'.

Owing to the increased use and popularity of the PCM as an analysis framework, and the potential limitations of using other methods that were developed for different theoretical perspectives, it is argued here that the PCM warrants its own structured method to enable its continued use and application. The PCM has been applied across many domains, including road, rail and aviation, and for many purposes, from accident analyses (Plant and Stanton, 2012a; Salmon et al., 2013a) to modelling situation awareness (Salmon et al., 2013b, 2014a); for its potential use to be fully realised, a formal method is required. In response to this, the research presented in this chapter details the development of a data-driven cognitive probe method that is more applicable to extracting the interaction between internal schemata and external world information, as epitomised in the PCM, in order that perceptual cycle processes can be better understood.

7.2 METHOD I: SWARM DEVELOPMENT

7.2.1 Data Collection

Critical incident data were collected from 20 rotary wing pilots. The full procedure for this is detailed in Chapters 5 and 6. For a detailed description of the participants and procedure, refer to Sections 5.2.2 and 5.2.3.

7.2.2 Data Analysis

In accordance with the guidelines on qualitative data analysis, the text was chunked into meaningful segments of approximately one sentence or less in length. This resulted in 904 text segments across the 20 interviews. Both deductive and inductive thematic analysis was used to analyse the data (as described in Section 6.2.2).

7.2.3 SWARM Development

The categories from the PCM taxonomy were used to devise the SWARM, a repository of cognitive prompts that are intended as a data elicitation method to help understand perceptual cycle processes. The probes for each category are presented in the SWARM repository of prompts provided in Section A.1. The prompts are designed to enable operators to reflect on the three elements of the PCM during decision making. The schema prompts (Figure 7.1) allow operators to reflect on the role of past experiences and expectations, for example whether they had experienced the situation before and whether previously held expectations influenced the decision-making process. The action prompts (Figure 7.2) aim to elicit information about different actions undertaken during an event, and the world prompts (Figure 7.3) aim to gather information about the information that was gained from the external environment and how this influenced the decision-making process.

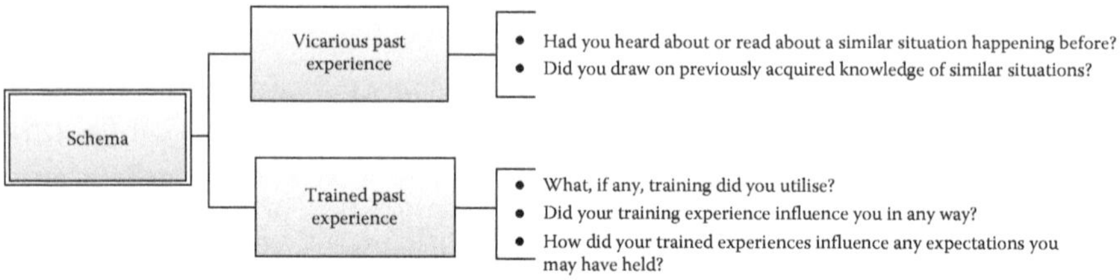

FIGURE 7.1 Example of Schema World Action Research Method prompts for the schema categories.

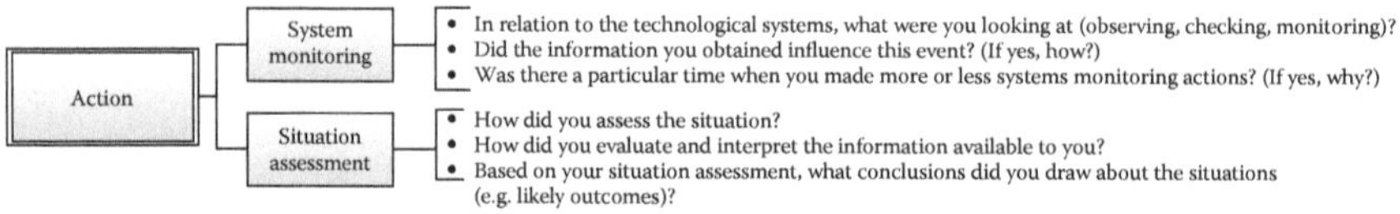

FIGURE 7.2 Example of Schema World Action Research Method prompts for the action categories.

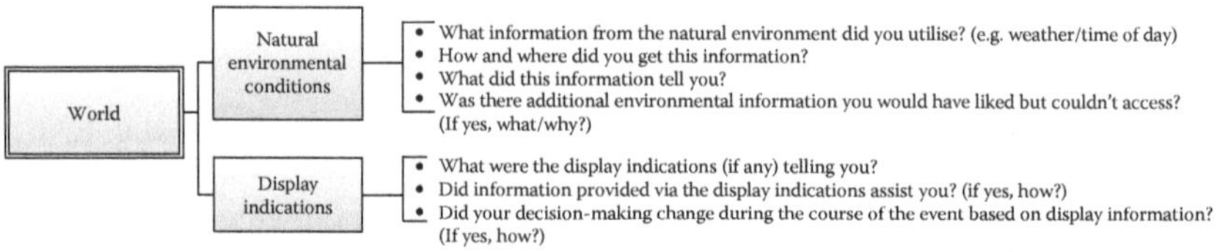

FIGURE 7.3 Example of Schema World Action Research Method prompts for the world categories.

7.2.4 Procedural Guidance

A total of 95 prompts have been included in the SWARM repository (Section A.1), and it is clearly not practical to ask all of the prompts in one interview session. In a pilot study conducted with another researcher, it took over two hours to ask all of the prompts (with answers), and this was for a relatively simple event (a driving journey). Time with subject matter experts is often limited, and therefore, SWARM needs to strike the balance of gathering detailed qualitative data but in a realistic time frame. In Chapter 6, the PCM taxonomy (which is the basis of the SWARM prompts), in combination with frequency analysis, was used to define the relative importance of the different PCM categories in different phases of dealing with aviation critical incidents. These data were used to define the SWARM interview schedule provided in Section A.2; that is, the interview schedule is structured around the top five PCM categories (and associated probes) relevant to the specific phases of dealing with an aviation critical incident. Therefore, only the most relevant probes are asked, but the interviewer has the option of asking additional probes if they are deemed relevant. Furthermore, if SWARM was utilised in different domains or for a different purpose, the SWARM prompts might be different. To assist with this, the interview schedule provides the PCM categories in order of importance for the next 10 most important subtypes (see Section A.2).

Section A.2 details the SWARM interview schedule, and the procedure is summarised in Figure 7.4. The first step in the process is to outline the purpose of the SWARM interview schedule and provide a high-level overview of the theory behind the PCM and the subtypes of the model. At each stage of the procedure, it is necessary to check that the interviewee is happy with the explanation provided and understands the nature of the process. The SWARM interview schedule commences with the interviewee providing a high-level overview of the event under analysis and then proceeds through the six phases of dealing with a critical incident, answering the prompts that have been identified. Owing to space constraints, only the high-level categories are detailed in the interview schedule; however, it is recommended that the full questions are listed in order before the interview commences to afford a smooth transition between questions. At the end of each phase, the interviewee is asked whether any other PCM categories are relevant to that phase. If the answer is yes, the interviewer asks the relevant questions from the repository. It should be noted that the interviewers who conducted the interviews for the study presented in this chapter had a couple of read-through of the repository to familiarise with the location of the prompts and then found it straightforward to find additional prompts when necessary. The interview ends once the relevant prompts for each phase have been answered.

7.3 METHOD II: SWARM TEST CASE

7.3.1 Participant

The participant was a 42-year-old, male, ex-airline pilot with over 10,000 hours of flying experience mainly acquired on B767s. The participant was interviewed about a critical incident he had been involved with in 2007 (engine failure on take-off) which occurred when flying a commercial passenger jet from a UK airport to the Mediterranean. The incident occurred when flying a B767, and at the time he had

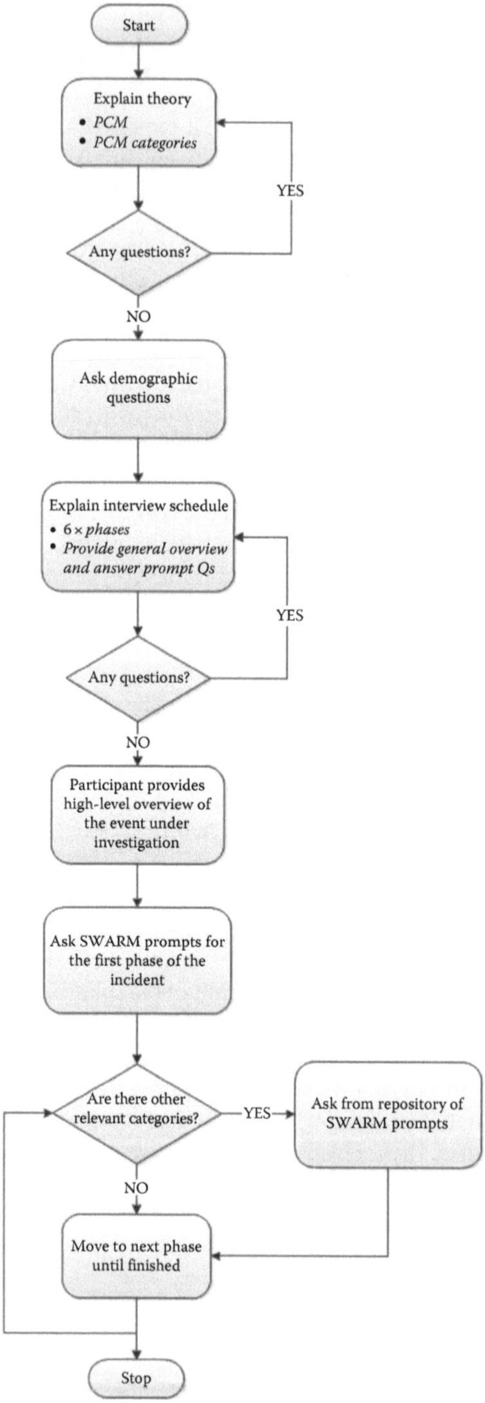

FIGURE 7.4 Schema World Action Research Method interview procedure.

approximately 8000 flying hours on type. Ethical approval for the study was granted from the Ethics Committee at the University of Southampton.

7.3.2 Data Collection

The SWARM interview was conducted twice with a six-week interval between the first and second interviews. A different researcher conducted the interview at time two, with the aim of assessing the test–retest reliability of the method. This is in line with the recommendations made in Chapter 4 when conducting a test–retest reliability study. The second interviewer was given a 20-minute training session which resembled the procedural guidance provided in Section 7.2.4 (i.e. introduce the purpose of the interview, provide an overview of the theory behind the PCM and the subtypes of the model, discuss the interview schedule and answer any questions that arose). The procedure outlined in Section 7.2.4 was followed: the participant provided a high-level overview of the critical incident and then the relevant SWARM prompts were asked for the six phases of dealing with a critical incident. After discussing each phase, the participant was shown the prompt sheet (i.e. the annotated PCM diagram on page 1 of the interview schedule) and asked whether any other of the PCM categories were relevant for that phase of incident. As procedural recommendation, the interviewer can use the prompt repository to describe the categories if asked for clarification from the interviewee. If any were identified, then these prompts were asked from the repository of prompts (provided in Section A.1). Each interview lasted for just over an hour (68 and 72 minutes, respectively). The interviews were audio-recorded and later transcribed.

7.3.3 Data Analysis

To align with the research aims of assessing both the theoretical validity and test–retest reliability of SWARM, the data were analysed in two ways. First, the transcripts were thematically analysed against the three categories of the PCM – schema, action and world – in order to understand the perceptual cycle processes involved in dealing with the critical incident. Second, content analysis was performed on each interview using the Leximancer™ text analysis software. Leximancer automates network creation by analysing verbal transcripts to identify concepts and relationships between concepts. The software uses algorithms linked to a built-in thesaurus and focuses on features within the transcripts such as word proximity, quantity and salience. Leximancer uses a weighting procedure to classify frequently appearing words as concepts and then determines how concepts are related to one another by measuring the co-occurrence of concepts (see Salmon et al., 2013b; Young et al., 2015 for more detail about the process). The output is a network representing concepts and the relationships between them. The concepts are of interest in this analysis because the automatically generated concepts were used to objectively compare the data sets from the two SWARM interviews. The only edits made to the Leximancer default settings was the margining of similar or plural words (e.g. decision/decisions, fly/flying, land/landing, engine/engines). The two interviews were analysed with Leximancer for each phase of flight and also as a whole, resulting in seven analyses.

7.4 RESULTS

7.4.1 Incident Synopsis

The pilot interviewed was the First Officer on the flight in the Pilot Flying role. He was therefore responsible for the handling of the aircraft, whilst the captain, in the Pilot Not Flying role, was responsible for the systems monitoring and management and communications. What follows is a synopsis of the incident amalgamated from the SWARM interviews:

> It was a nice gin clear day, no wind and so a very routine nice day. And we were going somewhere nice and so everyone was in quite high spirits about where we were going… it was a better than average day. I knew the Captain that I was flying with which adds to the relaxed atmosphere. Everything, setting the aeroplane up, departing, getting the passengers on, we went on time, there were no delays. I took off and then you retract the gear, on the 767 the nose wheel comes up underneath the flight deck and spins down in the undercarriage bay, so you get quite a lot of vibration and noise associated with that right underneath you…as the wheels come up…about 500–600 feet, and then it starts to go away. My first recollection of the onset of something strange was that the vibration and noise didn't go away…then there was a thump like bang and a lurch to the right….We could see the [exhaust gas] temperature was off the clock, had gone through the upper limit and there had been a lurch and a bang and it was making a rumbling noise…we agreed we had a problem with that engine and she [Captain] retarded the thrust lever on that side to see if she could get the temperature back down under control…we decided, that because of the banging and it was still vibrating even at idle, we would shut it down and all the time we carried on climbing and went through a thousand feet. Once we decided to shut the engine down then we were back into the routine of how it is taught in the simulator, once we had made that decision to keep the left one running and shut the right one down, the Captain made the PAN call to Air Traffic Control. I called for the engine fire checklist, she did it, the vital actions, from memory, which is thrust lever closed, fuel control switch to cut off, engine fire switch pull, with me cross checking her actions and then once she completed the vital actions we were at about two thousand feet and the procedure is a level acceleration on one engine, so we levelled off at two thousand feet. I retracted the flaps and once the flaps come in there is the after take-off check list to do, the rest of the engine fire checklist, so you memorise the first bit and you have to get the checklist out and do the rest of the actions. We asked for a turn to keep near the airport. The autopilot can go in once the flaps were up, so I was steering it on the autopilot and doing the thrust because you lose the automatic thrust when you are one engine, while she ran the checklist then once we'd finished doing those, once we had finished the checklist and the after take-off checklist so that the aeroplane is safe, at that point we switched roles. I became Pilot Not Flying…we returned to the airport and we met on the runway with the fire services.

7.4.2 Theoretical Validity of SWARM

The SWARM interview schedule was designed to elicit perceptual cycle processes, and the procedure was designed around gaining information that is most relevant at specific phases of dealing with a critical incident. Utilising the SWARM prompts elicited information from the three categories of the PCM (world information, actions undertaken and the schemata held). A summary of each phase is discussed in turn:

7.4.2.1 Pre-Incident Phase

The world SWARM prompts elicited information about the weather conditions, operational context and location. The pilot discussed the favourable weather conditions and the relaxed nature of that particular flight. In both interviews, the natural environment prompt of 'Was there additional environmental information you would have liked but didn't have access to?' elicited information about the tech-log which is a record of every previous flight on the aircraft. The pilots noted that on previous flights there had been some minor difficulties with starting the right engine. The pilot stated that at the time it was nothing more than a passing comment but he remembered it because it turned out to be significant. The actions at that phase were routine pre-flight procedures, and the schemata held were based on previous direct experience of pre-departure routines. Due to the prevailing conditions, the pilot held the expectation that it was going to be a better than normal day.

7.4.2.2 Onset of Problem

The world-specific prompts for the onset of problem phase specifically relate to physical cues, technological systems and display indications. Interestingly, the pilot had touched on all of these areas in the overview of the incident he provided at the start of this phase. This provides support for the content of the prompts in the different phases. Figure 7.5 presents the PCM annotated with the account of dealing with the critical incident from the onset of the problem phase through to the decision of shutting down the engine (the numbers denote the order of events). This is an amalgamated summary from the six phases of the incident. Unsurprisingly, world information alerted the pilots to the problem with the engine. The action prompt questions in this phase related to aviating but the pilot also touched on the issues of diagnosing the problem so the 'concurrent diagnostic actions' prompts were also asked. In relation to the 'direct past experience' prompts, the pilot stated that he had not experienced the situation before but mentioned that he had trained for it. Therefore, the 'trained past experience' prompts were asked. Interestingly, the pilot stated that the initial diagnosis was hard because the engine did not fail like it did in the simulator; this is exemplified in the following text extract: *from the point of making the decision to shut the engine down it was exactly like the simulator, before that it wasn't like the simulator and that is why I think it was difficult, it wasn't just following your training in the simulator, it took a bit of analysis.*

7.4.2.3 Immediate Actions

The immediate actions involved the captain throttling back the engine to idle, but when the vibration returned the pilots discussed whether to shut the engine down (as depicted in Figure 7.5). The pilot was PF, and therefore, manually flying the aircraft, he continued to climb the aircraft until 2000 feet where they levelled off. The pilot stated that there was no concern about the immediate safety of the flight because they were climbing and knew there was nothing around them to hit (i.e. the incident had not happened in a mountainous region) and the aircraft was under full control. The schema prompts elicited the fact that the pilot held the expectation, developed from training, that if an engine failed on take-off it would be more difficult to handle the aircraft. In the 'engine failure' training scenario, the aircraft is hard to handle

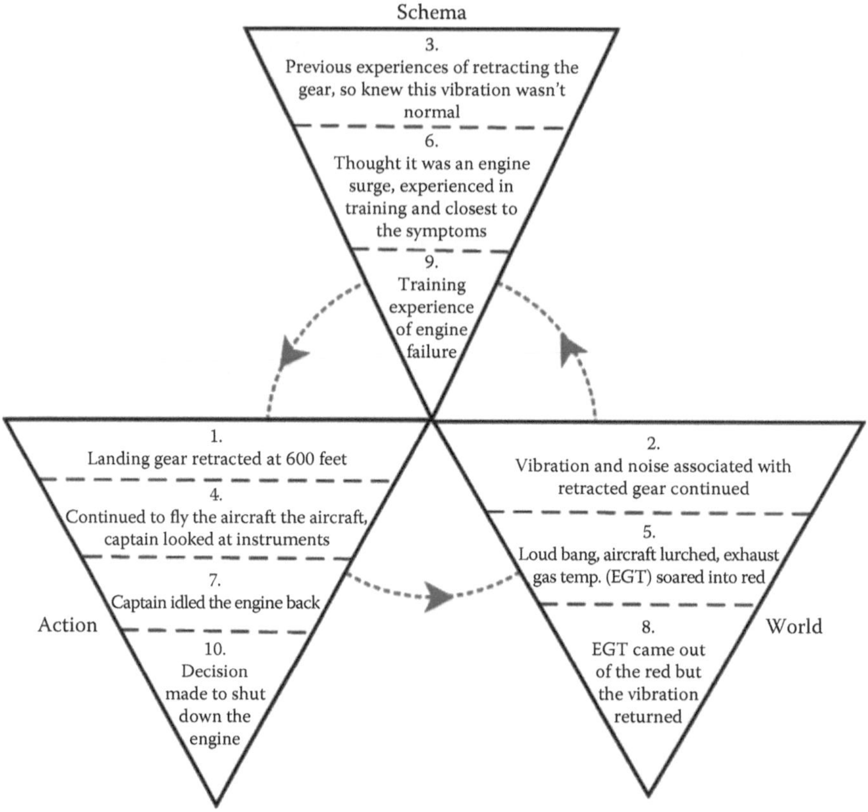

FIGURE 7.5 Perceptual cycle model annotated with extracts from the Schema World Action Research Method interview for the decision to shut down the engine (numbers denote the order in which the extracts should be read).

but the problem is easier to diagnose. In reality, the pilot felt the problem was harder to diagnose, that is, not as clear-cut as in training, whereas the aircraft handling was much easier.

7.4.2.4 Decision-Making

The primary decision that was made was to fully shut down the engine (rather than keep it at idle). This decision was based on world information (the continued vibration, the aircraft had lurched to the right and displays were showing a very high exhaust gas temperature) which the pilots knew from previous experience and training was not normal at that phase of flight (see Figure 7.5). During this phase, the aircraft had reached about 1200 feet and a secondary decision was made to keep climbing the aircraft to 2000 feet. The decision to shut the engine down resulted in the PF calling for the engine fire checklists and the captain worked through the checklist. During this phase, the pilot's primary action was to fly the aircraft, but he was also cross-checking the captain to ensure the correct engine was shut down.

The pilot described this as the most critical point, and he stated *all I was thinking is has she got her hand over the correct switch.* With the action prompts, the pilot described *...30 seconds of thinking what was the best course of action to take*; this led to the 'situation assessment' prompts also being asked in this phase.

7.4.2.5 Subsequent Actions

The subsequent action phase involved preparing the aircraft for a single-engine approach and landing. The pilots changed roles, but the airline flight procedures dictate that the pilot not flying (PNF) flies the approach for the pilot flying (PF) to land off; therefore, even though the roles had changed, the current PF kept flying to set up the approach. Other actions in this phase included the captain briefing the cabin crew of the situation and addressing the passengers, whilst the handling pilot re-programmed the computers back to the airport, and completed the landing checklist. The captain took over control of the aircraft at about 1500 feet and the interviewed pilot took over the monitoring and communicating role of the PNF. At this phase, the pilot was communicating with air traffic control. The 'declarative schema' prompts relevant to this phase confirmed that much of the process was following known standard operating procedures and that he knew the correct engine had been shut down because they still had power and the abnormal indications had gone.

7.4.2.6 Incident Containment

In the final phase, the pilot stated that they knew it would be a fast and hard landing because they were over maximum landing weight, and it would result in the brakes getting hot, which is something that the pilots had discussed. The captain landed the aircraft and over the radio the PNF spoke to the fire services who had met them on the runway. The fire services used thermal imagining cameras to check for signs of fire, and this provided world information to the pilots that there were no positive signs of fire and confirmed that the brakes were hot. The pilots were watching the brake temperature on the instruments because if they are over a certain level, taxiing is not an option as the brakes have the potential to melt the tyres. The brake temperature reduced to normal limits, and so they were able to taxi to a stand rather than disembark on the runway.

7.4.3 Test–Retest Reliability of SWARM

A key attribute of any method must be that it is reliable (Stanton and Young, 1999); that is, the method provides a consistent measure of what it intends to measure (or elicit). A method that is assessed to be reliable can be applied in a variety of circumstances, with confidence that consistent, and therefore comparable, data are being collected. A common way to assess reliability is to conduct a test–retest study to determine whether similar data are collected on two different occasions (as done here).

Table 7.1 presents the results from the Leximancer analysis of the two SWARM interviews. The results show the top five concepts for each phase of the interview and for the incident as a whole. Concepts are defined by Leximancer as collections of words that 'travel together' throughout the text. These are weighted according to their frequency of occurrence in concept containing sentences and non-concept

TABLE 7.1
Comparison of the Top 5 Concepts Generated by Leximancer for Interviews 1 and 2

	Whole Incident	Pre-Incident	Onset of Problem	Immediate Actions	Decision-Making	Subsequent Actions	Incident Containment
Interview 1	Engine(s) (100)	Engine (100)	Lurch (23)	Engine (100)	Down (100)	Flying (100)	Brakes (100)
	Down (64)	Nice (91)	Temperature (78)	Down (83)	Climbing (77)	Flaps (94)	Hot (85)
	Land/landing (42)	Easy (73)	Engine (57)	Fire (42)	Engine (50)	Engine(s) (88)	Down (85)
	Decision(s) (38)	Clear (64)	Down (57)	Wrong (33)	Land (41)	Land(ing) (75)	Fire (46)
	Time (31)	Day (64)	Thump (43)	Correct (25)	Left (36)	Crew (62)	Fast (38)
Interview 2	Engine (100)	Day (100)	Engine (100)	Flying (100)	Engine (100)	Land(ing) (79)	Engine (100)
	Down (66)	Aeroplane (100)	Vibration (50)	Engine (55)	Fails (89)	Approach (54)	Brake(s) (82)
	Land/landing (45)	Engine (94)	Flying (42)	Action (55)	Decision(s) (89)	Down (50)	Down (55)
	Decision(s) (43)	Problems (62)	Temperature (42)	Moved (45)	Producing (84)	Brief(ing) (50)	Approach (55)
	Time (39)	Nice (44)	Problem (38)	Fuel (45)	Down (68)	Cabin (46)	Runway (36)

Numbers in parentheses denote percentage relevance.

sentences. The concepts are assigned a percentage relevance number (100% being the most frequently occurring concept). The whole incident analysis shows that the top five concepts were the same in the first and second interviews; furthermore, these concepts occurred in the same order of importance (engine[s], down, land/landing, decision[s], time). These concepts reflect the nature of the critical incident, that is, an engine failure where a decision was made to shut down the engine and immediately land. The individual analysis revealed that across the 6 phases 11 concepts in the first interview were the same in the second interview (37%) and the order of importance varied across the interviews. Phase one (pre-incident) had the greatest match with 'engine', 'nice' and 'day' identified as top five concepts in both interviews. Phase three (immediate actions) and phase five (subsequent actions) had one shared concepts in the two interviews, 'engine' and 'landing', respectively. However, where concepts were different, an inspection of the transcripts revealed that very similar events were being discussed even if the terminology varied slightly. For example, in the first interview during the pre-incident explanation, the pilot made reference to how 'easy' the day was (i.e. nice weather, no delays, a familiar crew), but in the second interview, 'problems' was defined as a concept. A review of the transcript showed that this was in the context of *there were no problems with the aeroplane, …no technical problems with baggage loading* and is therefore akin to the 'easy' concept defined in the first interview. Other similarities exist, for example, in the first interview during phase five (subsequent actions), 'crew' is defined as a concept, whereas in the second interview, 'briefing' and 'cabin' are defined as concepts, but the text is referring to the same event of briefing the crew and preparing the cabin for an immediate landing.

7.5 DISCUSSION

SWARM is a semi-structured interview schedule based on the increasingly utilised PCM. Its unique contribution is to elicit data in order to examine decision-making in more detail using a set of cognitive probes developed from the PCM. The development of SWARM was data-driven as it evolved from a detailed PCM taxonomy. The prompts are designed to allow the interviewer to probe for insights into the taxonomy components. The goal is not to find out whether each component is present in the event but to uncover the nature of each component in terms of how it influenced the decision-making process. SWARM was applied to explore an aviation critical incident. The data were explored in terms of the theoretical validity and the test–retest reliability of the method.

7.5.1 Validity and Reliability of SWARM

There is no shortage of methods in Ergonomics; however, the extent to which they are based on academic theory or verifiable evidence is questionable (Stanton, 2002; Stanton et al., 2013). Furthermore, Salmon et al. (2009) argued that reliability and validity should be critical considerations when selecting methods, although often little is done to assess this (Stanton et al., 2013). Methods can be reliable (provide consistent measurement) without being valid (provide an accurate reflection of a phenomenon), but they cannot be valid without being reliable (Stanton and Young, 1998, 1999). Annett (2002) differentiated between analytical and evaluative methods. SWARM is an analytical method because it is designed to help the analyst gain an understanding of mechanisms (i.e. perceptual cycle processes) underlying humans and machines. Stanton et al. (2013) argued that analytical methods need to stand up to construct validity, that is, how acceptable the underlying theory is. Chapter 5 assessed the construct, or theoretical, validity of the PCM. Furthermore, in recent years there has been resurgence in using the PCM as a theoretical framework to model human–world interactions in safety-related research (Stanton and Walker, 2011; Salmon et al., 2013a). This suggests that researchers are fundamentally comfortable with its representation of information processing. SWARM is therefore founded on strong theoretical principles and so conforms to Stanton et al.'s (2013) definition of construct validity. Furthermore, the analysis presented in Section 7.4.2 demonstrates that data elicited from SWARM allow an understanding of perceptual cycle processes to be gained, suggesting that SWARM is a valid method for eliciting perceptual cycle information.

In the test–retest reliability analysis of the individual phases, 11 concepts in the first interview were the same as in the second interview (this was out of a total of 30 possible pairings). On the surface, 37% agreement may appear relatively low; however, a review of the transcripts revealed that when concepts differed, the text relating to these concepts was referring to the same information or events of the incident. Leximancer was chosen to perform the test–retest reliability assessment because the automatically generated concepts are objectively determined compared to a human analyst. The test–retest reliability study presented in Chapter 4 provided a manual analysis of a CDM interview, also for an aviation critical incident. Here, the constant comparison technique was used to define high-level themes within the two interviews (examples of themes include environmental conditions, the problem and location information). If this approach had been employed in this analysis, it is likely that there would have been a very high level of agreement between the two interviews because the transcripts are fundamentally the same, although the terminology varied which resulted in Leximancer defining different concepts. This is confirmed by the fact that the whole incident analysis revealed the same top five concepts in the same order for the two SWARM interviews (see Table 7.1). This suggests that overall the test–retest reliability of SWARM is high, that is, on different occasions, the SWARM prompts will enable the same information to be elicited. In light of the findings presented here, other researchers are recommended to explore a mixed approach for assessing inter-rater reliability. Leximancer provides an objective and relatively quick approach, although it may produce misleading results due to its sensitivity at defining concepts. A manual analysis takes longer and is open to criticism about its subjective nature; however, it allows for a project-relevant analysis to be conducted. A mixed approach could involve using Leximancer for the whole interview data and a manual approach some selected phases.

7.5.2 Theoretical and Practical Applications

SWARM has provisionally been demonstrated to be a valid and reliable method, and as such, we would recommend its use for eliciting perceptual cycle processes.

In the introduction, other approaches to achieve this were described, such as verbal protocol analysis and thematically analysing CDM data or documents like accident reports. We would argue that due to the data-driven development of SWARM, it provides a semi-structured approach to tap into the relevant elements of the PCM, giving equal weighting to schemata, actions and world information, in which other interview schedules, like the CDM, fail to do. SWARM is not directly comparable to verbal protocol analysis as they are different data elicitation techniques, the former being a retrospective interview and the latter being a concurrent think-aloud approach. There are known limitations with verbal reports with regard to whether they change the nature of task performance or capture all information (Bainbridge, 1999). In a study of simulated driving behaviour, Banks et al. (2014) used a mixed approach whereby verbal reports were supplemented with post-task CDM probe questions. The SWARM prompts could be utilised in the same way if the purpose of the analysis was to capture PCM information (e.g. as in Salmon et al., 2014), with the intention that this will provide information related to the PCM that was not verbalised.

Stanton (2005) argued that method development comes from processes based upon enhancements, modifications and combinations of existing methods. There is a long history of this in relation to the CDM as this was a descendent of Flanagan's (1954) original Critical Incident Technique. The probes were then enhanced by Ericsson and Smith (1992) to gain additional information about cognitive structures and processes that appear to mediate expertise. The CDM is clearly a predecessor of the SWARM insomuch that SWARM also provides list of cognitive probes with the aim of eliciting reasons and strategies for decision-making. The research presented in other chapters in this book has advocated the use of the CDM to uncover perceptual cycle processes because the CDM elicits knowledge for decisions that are devised through the recognition of critical information from the environment (i.e. world) based on past experience and expertise (i.e. schemata). However, the development of the CDM was based on the theoretical underpinnings of the RPD model of naturalistic decision-making, which primarily describes the decision-making process inside of the head of the decision maker. The PCM has been advocated as a more distributed view of decision-making owing to its inclusion of internally held schemata and externally presented environmental information as influencing factors in the decision actions that are undertaken. O'Hare et al. (1998) argued that the importance of theoretically driven CTA should not be underestimated as this has a huge potential to impact on subsequent system design and training improvements. Therefore, it seems appropriate to develop a method in line with this theoretical perspective. Furthermore, the CDM has been criticised for not being specific enough. For example, O'Hare et al. (1998) utilised the CDM to investigate cognitive information relevant to white-water rafting guides and their biggest criticism of the CDM related to the schematic side of decision-making. The CDM probes did not allow the authors to establish whether expert guides were using direct analogies of previous experience or a prototype matching approach. Additionally, their findings showed that expert raft guiders took into account a wide range of external cues for determining their course of action, but the CDM only asks one high-level question in relation to cues. The data-driven development of the SWARM prompts from the PCM taxonomy has ensured that detailed and specific prompts were derived, and therefore, SWARM has the potential to elicit much more information in relation to schematic processing and the environmental information that is attended to.

It is envisaged that the SWARM will be utilised in a variety of research settings. Its most common application is likely to be as a semi-structured interview schedule for decision-making situations, with the intention that it will aid operators in articulating knowledge that is difficult to verbalise. Unlike the CDM, the SWARM is also intended for use for non-critical decision-making situations and the prompts are written to reflect this. As argued by Rodrigues de Carvalho et al. (2009), research usually focuses on the extremes of situations, but what is less well researched are normal conditions, with normal operators using normal equipment. Incidents and accidents are the by-product of this normal functioning with people acting in a way that made sense to them at the time (Dekker, 2006). The only category of SWARM that explicitly applies to dealing with a critical incident is 'problem severity', and this can be omitted in the prompt selection phase of the procedure. The SWARM repository provides researchers with a comprehensive list of probes that have the potential to tap into key areas of decision-making and can be utilised in a variety of formats to suit the researchers' needs. Previously, to understand the perceptual cycle process in data collected via the CDM, the analysis required deductive thematic coding based on the high-level PCM categories (e.g. see Plant and Stanton, 2013a). Data elicited from the SWARM will already be structured around these high-level categories, and therefore, value lies in the time saved by removing a step in the analysis process.

7.5.3 Future Research

SWARM was developed using a structured and exhaustive approach which produced a comprehensive list of prompts. However, it is only through correlating evidence with other sources that continued confidence in the validity and reliability of the method can be established. Detailed procedural guidance has been provided and is designed to assist others in this pursuit. Although the prompts have been written from the context of aeronautical decision-making, only two of the prompt categories (aviate and aircraft status) make explicit reference to aviation terminology. With only minimal modifications, it is envisaged that the prompts can become a domain-independent research tool.

In its current format, SWARM is developed for examining decision-making when dealing with a critical incident, and a selective approach is advocated so that the most prompts are chosen for different phases of dealing with a critical incident. This is in line with Militello and Hutton's (1998) assertion that when conducting CTA using probes it is not expected that all probes will be equally relevant for all situations or domains. In the procedural guidance, interviewers are encouraged to ask interviewees if any of the other prompts are relevant to their current discussion or use their own judgement in deciding whether additional prompts should be asked. This was evident in the case study interview when additional prompts (e.g. concurrent diagnostics, trained past experience and situation assessment) were asked due to the information that had been elicited. Both of the interviewers asked additional prompts, and interviewer 2 was not familiar with the development of SWARM and had only been introduced to it in the pilot study. This demonstrates that with minimal training an interviewer can become familiar enough with the repository of prompts to be able to deviate from the procedural script.

The theoretical underpinning of the SWARM also means it has the potential to be more applicable to understanding non-expert decision-making processes. The RPD model and associated CDM were developed to explain and understand expert decision-making, but the PCM, however, thus the associated SWARM, is non-specific in terms of expertise. Schemata are developed through past experience and expectations, and therefore, expert decision makers will have a larger repertoire of schemata to draw upon to enable them to select relevant environmental cues and thus be more selective than novices in the information they attend to. It is anticipated that through the application of the specific prompts in relation to these key areas (e.g. past experiences, environmental information), the SWARM will be able to provide data that allow different levels of expertise to be usefully compared, although it is only with further research of this nature that this will be known for certain.

7.6 CONCLUSION

The PCM provides a process-orientated approach for understanding how internally held mental schemata interact with information perceived in the external environment to produce actions and behaviours. With its description of the reciprocal relationship that exists between the mind and the environment, the PCM allows insights to be gained about local rationality and the reasons for making decisions. To date, the application and utility of the PCM has been limited by the lack of a structured methodological approach for eliciting perceptual cycle information, hence the motivation behind this work. We have produced PCM-based cognitive prompts designed to elicit data about decision-making from the perspective of the three elements of the PCM. It is essential to establish the validity and reliability of methods so that they can be used with confidence (Stanton and Young, 1998; Salmon et al., 2009). The case study data have taken the first steps at establishing the validity and reliability of the method, and detailed procedural guidance is provided to assist other researchers with this pursuit. The research presented so far has focused on individual decision-making; however, socio-technical systems are characterised by individuals working as part of a team, and therefore, Chapter 8 explores the utility of applying the PCM to explain team critical decision-making processes.

8 Team Perceptual Cycle Processes

8.1 INTRODUCTION

So far, the ideas explored in the book have been from the perspective of an individual operator. The purpose of this chapter is to apply Neisser's (1976) perceptual cycle model (PCM) to a team activity and explore how the approach translates. The PCM has traditionally been concerned with the individual level of analysis; however, this is not as relevant in the study of modern, complex socio-technical systems that are characterised by teamwork. By way of a case study method, the chapter investigates decision-making processes of a search and rescue (SAR) helicopter team when dealing with a critical incident. Traditional perceptual cycle representations and novel network-based representations are compared and contrasted for different phases of the critical incident. A distributed cognition perspective of the PCM applied to SAR team decision-making requires a shift from the traditional notion of the PCM that focuses on the individual to one that focuses on the team. The network approach has been successfully applied to investigate distributed cognition in a variety of domains including energy distribution (Salmon et al., 2008), air traffic control (Walker et al., 2010), submariner decision-making (Stanton, 2014) and driving (Salmon et al., 2014). However, traditional network analysis does not explicitly consider the interaction between a person and the world and how this impacts decision-making.

8.1.1 PCM, Decision-Making and Teams

The relationship between naturalistic decision-making (NDM) and the PCM (see Figure 3.1) has been extensively described in the previous chapters; the discussion here focuses on comparing the PCM with the more dominate NDM model and Klein's (1998) recognition primed decision (RPD) model and discusses the PCM in the context of teams. NDM refers to the process of how people use their experience to make decisions in the real world as opposed to laboratory settings (Klein, 1998). NDM is a way of understanding how people handle confusions and pressures in their environment which is usually characterised by limited time, goal conflicts and dynamic conditions (Klein et al., 1989). The focus of study in this domain is described by Stanton et al. (2009, p. 482):

> [as] purposeful behaviour, i.e. teams working together on tasks towards some end goal in highly dynamic collaborative environments.

Klein (1998) highlighted dynamic conditions, that is the changing situation, as one of the key features of NDM. As new information is received, or old information becomes invalid, the situation and goals can be radically transformed. The most popular model in the NDM domain is Klein's (1998) RPD model. In summary, this captures how experts make decisions based on recognition of past experiences that are similar to the current situation. These experiences are used to generate a workable option in a process known as satisficing (Klein, 1998). In complex cases, evaluation of the option reveals flaws that require modification or the option is rejected in favour of the next most typical reaction. The RPD model considers mental simulation as the mechanism for the cyclical nature of decision-making, but this is only internal to the decision maker. As such, the cyclical nature of a changing external environment is not fully captured in the RPD model. Similarly, the implementation of the model does not connect the internal process of the decision maker to the external environment in which decisions are made. Therefore, the explanation provided by the RPD model is primarily one of the decision-making processes occurring in the head of the decision maker.

Klein's (1998) discussion of team decision-making follows a similar vein and focuses very much on the 'team mind', stating that team cognition can be inferred from three sources: the team's behaviours, the contents of the team's collective consciousness and the team's pre-consciousness. The role of the environment, such as artefacts available or the influence of other team members on their colleagues, is not acknowledged. Whereas decision-making of any kind, especially in the dynamic conditions that characterise the NDM environment, is a product of the interaction of the processes going on in the head of the decision maker and the conditions in the external environment. As Dekker (2006) argued, in order to truly understand decision-making, it is essential to account for why the actions and assessments undertaken by a decision maker made sense to them at the time. This will be based on internal information in the head *and* external information in the environment. As such, it is proposed here that Neisser's (1976) PCM is a more suitable framework to model decision-making processes because it accounts for the cyclical interaction that occurs between an operator and their environment in a way that is not fully captured by the RPD model.

Carvalho et al. (2005) have argued that feedback and feed-forward loops are one of the defining characteristics of NDM, where dynamic situations are characterised by a series of events taking place overtime. Schemata held by individuals guide perceptual exploration in the world as they prime a perceiver in terms of what to expect from the environment based on past experiences; this is the feed-forward element of the PCM, where early actions are taken using anticipatory strategies (Carvalho et al., 2005). Interaction in the world results in schemata being modified by information attended to in the environment, which in turn modifies further environmental interaction (i.e. feedback is derived from the environment). Klein (1998) attempted to incorporate the role of schemata into the RPD model via the use of expectancies. However, Lipshitz and Shaul (1997) argued that the original RPD model was too simplistic in its representation of schemata. As such, they proposed a schema-driven RPD model which is similar to the PCM insofar as schemata are viewed as the drivers of situational information available to the decision maker. Unlike the PCM, however, the schema-driven RPD model does not include feedback loops and as such, like the original RPD model, overlooks the modifying effect that the environment can have on the decision maker's schemata. The representation of both feedback and feed-forward loops in the PCM adds weight to the justification of using it to model NDM.

Despite the widespread application of the PCM, detailed throughout this book (see Chapter 2 for a review), the framework has not been explicitly applied to team decision-making, which is highly relevant given the prevalence of teamwork in Ergonomics domains. This is supported by Hollnagel's (1993) argument that team behaviour should be analysed at a macro level, that is, taking into account the environment and context in which the team interact. The PCM can conceivably be used to explain team processes in decision-making situations; by linking the mind to action to world back to mind, the PCM has the potential to allow analysis and interpretation to transcend the individual level of a system.

8.1.2 Networks and the Analysis of Distributed Cognition

An overview of distributed cognition is provided in Chapter 1. Traditionally, ethnography has been the method through which distributed cognition has been understood (see Hutchins, 1995a). Walker et al. (2010) described ethnography as a form of naturalistic, qualitative description based on observation, usually in the form of participatory observation. The undertaking and outputs of the ethnographic approach can be time consuming, incompatible with objectivity and validity in measurement and not always amenable to generalisation (Walker et al., 2010). Walker et al. (2010) warned that these problems limit the practical value of distributed cognition and serve as a barrier to what is potentially a useful approach to the analysis of complex systems.

To overcome the challenges of modelling distributed cognition, Stanton and colleagues devised event analysis of systemic teamwork (EAST; Stanton et al., 2008) as a suitable method to do this. EAST is underpinned by the notion that complex collaborative systems can be meaningfully understood through a network-of-networks approach. Walker et al. (2010) employed EAST to represent distributed cognition in air traffic control and, in doing so, argued that the method puts ergonomics analysis in touch with the distributed cognition perspective, and as such, the output was much more traceable than comparative ethnographic techniques. The original version of EAST required input from a number of data sources, including hierarchal task analysis, critical decision method, coordination demand analysis, communications usage diagram and operation sequence diagrams (Stanton et al., 2013). However, Stanton (2014) presented a shortened version of EAST in which networks are developed directly from raw data; the approach used here has been inspired by the shortened version of the EAST methodology (see Chapter 9 for a more detailed description about EAST).

Networked-based representations have been used within the theory of distributed situation awareness (DSA), devised by Stanton and colleagues and underpinned by the PCM; DSA was primarily created to account for situation awareness (SA) in teams. This considers SA as an emergent property arising from people's interaction in the world, where the mind is situated in an interdependent relationship in the world (Stanton et al., 2006; Salmon et al., 2009). This perspective sees teams holding compatible, rather than shared, SA as no two individuals could ever possess identical SA due to the unique nature of schemata, developed through past experiences and expectations (Stanton et al., 2009). Sorensen and Stanton (2011) supported this view when they found that team SA was expressed through methods that assessed DSA rather than methods that addressed individual (or psychological) SA. The DSA approach has been shown to account for team SA (Sorensen and Stanton, 2011) and DSA is underpinned by the PCM (amongst other theoretical perspectives); it therefore makes sense that a team perceptual cycle is explored. Klein (1998) argued that one of the key goals of NDM research should be to understand how people in teams make decisions. In most NDM domains, and aviation is no exception, decisions are made by individuals working in teams who need to coordinate and communicate with each other.

8.2 METHOD

8.2.1 Critical Decision Method

The CDM is described in detail in Chapter 4. The method was applied here to study a team by interviewing team members about a critical incident they had been involved with.

8.2.2 SAR Environment

SAR is the task of searching for missing persons or location and recovering persons in distress and delivering them to a place of safety. SAR missions usually occur in a coastal operating environment, although it is not limited to this as support can be given to lead searches and hospital transfers. In the United Kingdom, SAR activity is carried out by Her Majesty's Coastguard which is responsible for coordinating all civil SAR efforts (Baber et al., 2013). An airborne SAR capability is available 24 hours a day, 7 days a week, in all weather conditions. SAR is a multi-organisation operation, including the national coastguard, the local SAR authority (including local maritime rescue coordination centres and coastguard rescue teams), the Royal National Lifeboat Institute and the local airfield (including air traffic services and ground support). In addition to the aircraft crew (pilot flying [PF], pilot not flying [PNF], winch man [WM] and winch operator [WO]), coastguard services and air traffic services are integral to the SAR operating environment.

8.2.3 Participants

The four SAR crew members who were present in the helicopter at the time of the incident were PF, PNF, WO and WM. All four crew members were male. The PF was 48 years old with 5500 hours of experience (600 on type). The PNF was 55 years old with 9000 flying hours experience (1500 on type). The WO was 65 years old with 6000 hours of relevant experience (500 on type), and the WM was 60 years old with 7500 hours of relevant experience (850 on type). The crew were involved with a critical incident during a winch training exercise when flying a Sea King AW139. The incident involved a fault with the engine oil temperature (EOT) in one of the engines.

8.2.4 Procedure

The crew members were interviewed separately at their helicopter base. Normally, the first phase in a CDM interview is identifying a suitable incident to discuss, that is, a critical incident is defined to the participant as being 'a non-routine or unexpected event that was highly challenging and involved a high workload' and the participant chooses a suitable incident. However, here the incident had to be predefined to ensure the relevant crew were on duty at the time of the interviews. Therefore, prior to the interviews, the incident was predetermined with the PF, the other crew members knew they would be interviewed about a critical incident but did not know which one. This step was taken to avoid the crew members rehearsing the incident or discussing it with each other as this may have interfered with their recall. The crew were on duty at the time of the interviews; therefore, interview time could not be guaranteed and so the shortened version of the CDM was used; that is, only the timeline construction and deepening probe phases were conducted.

The nature of the interview was explained to each participant. Each participant provided a high-level overview of the incident and structured a timeline of events. After the incident description/timeline construction phase, the cognitive probes were asked in relation to the decision-making made during the incident. Ethical permission for this study was granted by the Research Ethics Committee at the University of Southampton. The interviews were audio-recorded and later transcribed. The interviews took place 6 months after the critical incident occurred.

8.2.5 Data Analysis

A similar method was employed that has been described in previous chapters. In accordance with guidelines on qualitative data analysis, the transcribed interviews were chunked into meaningful segments of approximately one sentence or less in length (Strauss and Corbin, 1990). Deductive thematic analysis was used to analyse the text segments using the high-level PCM coding scheme described in Chapter 4. Chapter 3 described various ways that coded incident data can be structured in order to aid explanation. Here, the data will be structured around an amalgamated timeline of events generated by the four crew members. The perceptual cycle representation of two of the phases (Section 8.3.2) is compared to the network representation (Section 8.3.3).

8.3 DATA INTERPRETATION

8.3.1 Incident Synopsis

The synopsis provided later is an amalgamated account from the four interviews provided by the crew members. Six key phases were amalgamated from the individual

TABLE 8.1
Phases of the Critical Incident

Phase	Description
Dummy run briefing	The briefing phase occurs during the dummy run which is a mock up of the winching exercise (without the WM being deployed) in order to assess the risks. Briefing is a two-way flow of information between the PF in the cockpit and the WO in the back. The PF is assessing the power of the aircraft and states what will be done in the event of an engine failure. The WO acts at the eyes for the PF, stating things that have the potential to affect the WM such as hazards on the vessel, radars, aerials and vessel movement. In preparation for the live phase, the aircraft descends to 40 feet above the vessel.
Live run	This is the same as the dummy run but the WM is winched out to height; in this exercise, the intention was to put him onto the deck of the vessel. The WO is responsible for the WM and needs to react to the dynamic situation, for example, if the WM is pushed back by the wind. The WM and PF maintain constant dialogue to let each other know what is happening.
Critical incident	Onset of emergency situation. Whilst the WM was over the vessel, the pilots in the cockpit were alerted to a problem with the EOT via a flashing amber caution light. The situation was communicated via the intercom to the WO in the back and the WM could hear this via his headset.
Immediate actions	The decision was made to abort the training and return the WM to the aircraft. Other potential options available to the crew were to leave the WM on the deck of the vessel and to cut the winch wire; however, these options were not necessary for the situation as there was time to bring the WM back in.
Diagnostics	Whilst the WM returned to the aircraft, the PF trickled the aircraft forward. Once the WM was on board, the WO called 'safe, safe, safe' and the PF transitioned the aircraft into a safe flight configuration (40–60 knots and up to 200 feet). The PNF utilised the flight reference cards to ensure they had carried out the correct procedure and had not missed any memory items. Once the aircraft was in a safe configuration, the pilots were able to assess the situation by looking at other indications to ensure that it was just an EOT problem and not a sign of something more serious, for example, that an engine failure was imminent.
Return to base	When the aircraft transitioned out of the winch hover, the EOT naturally decreased due to a flow of cooler air. This took the temperature reading from red to amber, which gave the aircraft 30 minutes flying time so the decision was made to return to base which was 7 minutes flight away.

Keys: EOT, Engine oil temperature; PF, pilot flying; PNF, pilot not flying; WM, winch man; WO, winch operator.

timelines and are used as the basis to structure the data analysis; these phases are summarised in Table 8.1.

> The incident occurred when flying an AW139 helicopter as part of a routine SAR winch training exercise over a vessel. The crew had completed the dummy run phase of the exercise and were in the live run phase over the vessel. The aircraft was hovering at

40 feet and the WM had been winched out of the aircraft and was over the vessel when the pilots were alerted to a problem with the EOT via a flashing amber caution light. The digital scale for the EOT had gone through the amber reading and into the red. The high temperature meant the crew had one minute of flight time before the engine had to be shut down. The WM had to be rapidly returned to the helicopter by the WO whilst the PF put the aircraft into a safe configuration and the PNF attempted to diagnose the problem using the flight reference cards. Transitioning from the hover into forward flight caused air to circulate and naturally cool the engine oil and the EOT indication dropped back into the amber zone which gave the pilots 30 minutes of flying time. The decision was made to return to base which was seven minutes away.

8.3.2 Perceptual Cycle Representation

For each phase of the incident, each crew member's data were coded as schema, action or world, and the text segments were modelled into the PCM. This method has been successfully applied to explain aeronautical decision-making from an individual perspective (e.g. Chapter 4). However, when using this to structure team data, limitations of the PCM representation became apparent; this will be exemplified using data from two of the phases of flight: live run and the critical incident (Figures 8.1 and 8.2 respectively).

8.3.2.1 Live Run Phase

This phase is the same as the dummy run, but the WM is winched out to height with the intention to place him onto the deck of the vessel. All crew members hold schemata for the situation based on training and past experience (PF_S, PNF_S, WM_S, WO_S in Figure 8.2), which allows them to perform their role-specific actions (PF_A, PNF_A, WM_A, WO_A). The WO is responsible for the WM and needs to react to the dynamic situation, such as if the WM is pushed back by the wind. The interactions between the different elements of the PCM and between different crew members are particularly evident when looking at the WO and PF interactions. During this time, the WO acts as the eyes for the PF and his actions (WO_A) become world information for the PF (PF_W). Coupled with the PF's schemata developed from previous training and experience, he is able to perform the role-specific actions required for this phase of flight. The benefit of representing data with the PCM is that the cyclical nature of information processing is clearly depicted, and this works well at the individual level. However, when looking at the model from a team perspective, the interconnectivity between the different actors (e.g. PF and WO) and the three elements of the PCM is harder to convey.

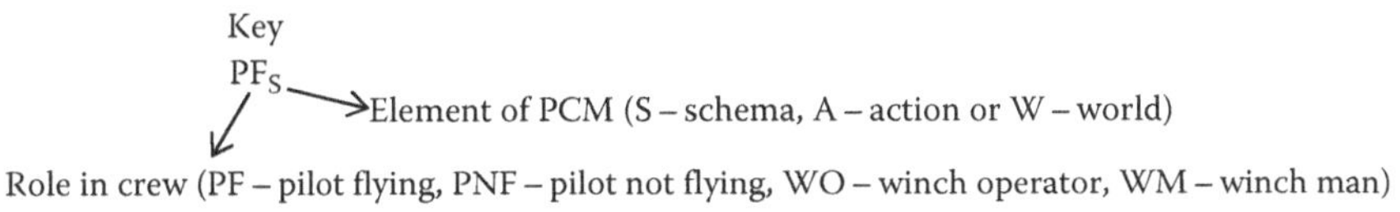

FIGURE 8.1 Key for the perceptual cycle representations.

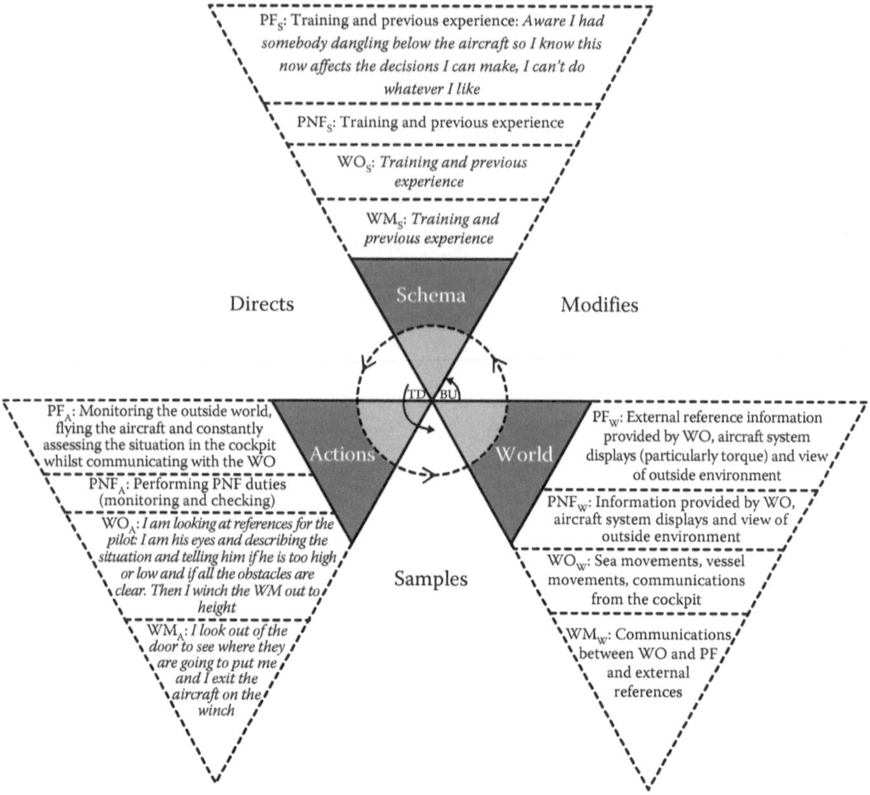

FIGURE 8.2 Perceptual cycle representation of phase two: live run.

8.3.2.2 Critical Incident Phase

Until the onset of the emergency situation, the crew were engaged in a routine training exercise and therefore the majority of their actions were based on schemata generated from training and past experiences and they performed routine actions. When the WM was over the vessel, the pilots in the cockpit were alerted to a problem with the EOT via a flashing amber caution light (PF_W, PNF_W in Figure 8.3). The situation was communicated via the intercom to the WO in the back, and this was heard by the WM in his headset. Interestingly, the two pilots held different schemata for the same situation. The PF had a schema developed from training in which the EOT displaying red upper limits meant they had 1 minute of flight time before the engine needed to be shut down (PF_S). However, the PNF had experienced the same event a few weeks before and therefore he held the schemata that the situation was likely to be spurious and was not critical (PNF_S). During the incident, the rear crew were kept informed about the situation from the pilots (WO_W, WM_W), and they held their own schemata for the situation based on their role-relevant experiences (WO_S, WM_S). Again, the cyclical nature of interaction is clearly demonstrated by this PCM representation (Figure 8.3), but the interconnectivity between the crew members is not

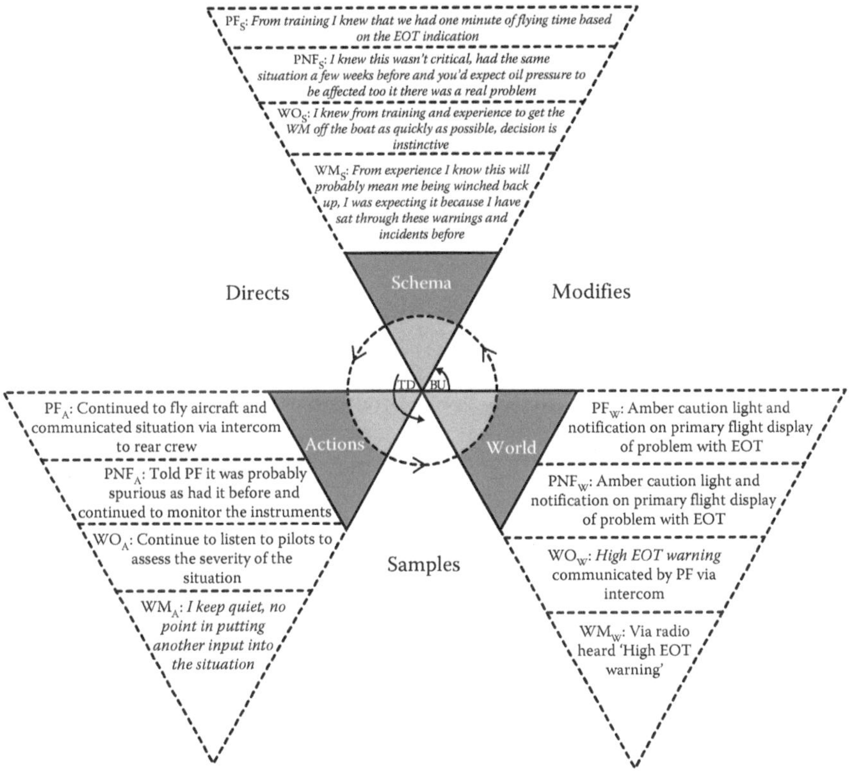

FIGURE 8.3 Perceptual cycle representation of phase three: critical incident.

adequately demonstrated. Furthermore, it is not immediately clear how much processing is top-down (schema-driven) or bottom-up (world-driven) or indeed if there is a difference. Owing to the limitations of using the PCM to represent team data, a network approach will be explored.

8.3.3 Network Representation

Network modelling has become a popular way of representing complex interactions including air traffic control (Walker et al., 2010), distributed cognition (Stanton, 2013) and SA (Salmon et al., 2013, 2014a,b) (Figure 8.4). Various programs exist to automate the process of network generation (e.g. Leximancer™); these have the advantage of removing analyst subjectivity, but the analysis is constrained by the

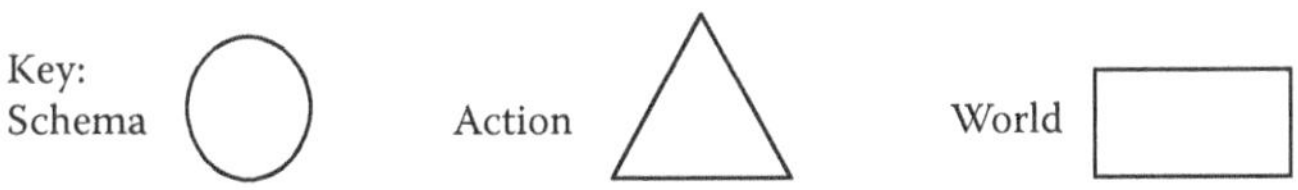

FIGURE 8.4 Key for the network representations.

limits of the software tool. For this analysis, the networks were required to depict the three elements of the PCM, so automated analysis tools were deemed unsuitable. The coding scheme used to code the data has previously been subjected to rigorous tests of reliability (see Chapters 4 and 5) which increases confidence in the reliability of this manual analysis. To enable direct comparison with the PCM representations of the data, the nodes for each network were derived from the text segments in the CDM data. Figures 8.5 through 8.10 depict the six phases of the critical incident via a network-based representation. The key features of these networks are as follows:

- The element of the perceptual cycle is represented by the different shapes of the network nodes (see key).
- Directional arrows indicate flow of information. Bidirectional arrows illustrate cyclical information processing (e.g. in Figure 8.5, the process of looking at 'external references' and 'communicating with the crew' is a continuous process of information gathering and dissemination).
- Crew members are depicted via role initial.

The benefit of a network-based analysis over the traditional PCM representation is that the interconnectivity between team members engaging in the perceptual cycle is illustrated and the individual perceptual cycles of each team member is not lost. For example, in the dummy run phase (Figure 8.5), the WO's information processing is both bottom-up driven via the external references available to him (sea and vessel movements and hazards) and also top-down driven via his training and previous role experience (schemata) that direct what he should be attending to in the world, hence the bidirectional arrow between these nodes. This then results in his actions of relaying information to the pilots (the perceptual cycle for the WO). Unlike the

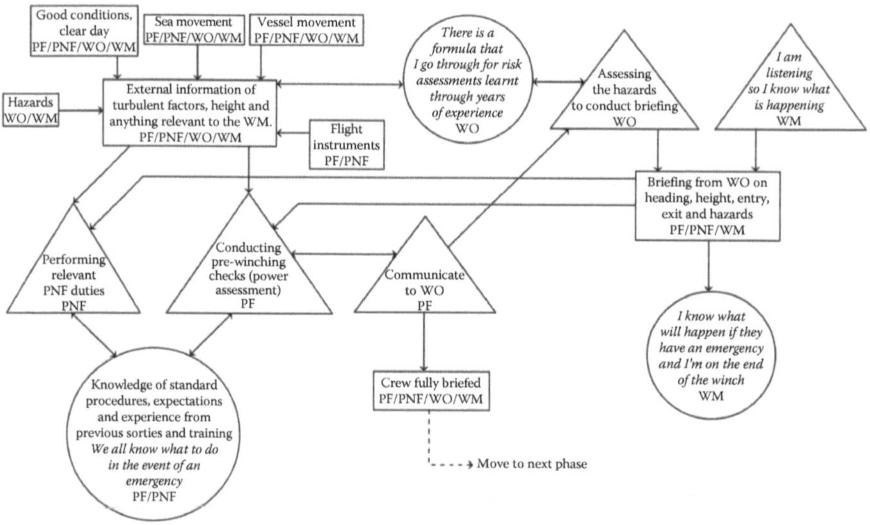

FIGURE 8.5 Network analysis of phase one: dummy run.

representation of the data in Figures 8.2 and 8.3, Figure 8.5 also illustrates the interconnectivity between crew members, for example the actions of the WO become world information for the other crew members. Furthermore, as demonstrated in Chapter 5, the data from each phase of incident were collated into a frequency table that captured 'from–to' links between the three categories of schema, action and world, and were recorded in a frequency count matrix table (see Table 8.2).

A further advantage of representing the CDM data as a perceptual cycle network is that automatic actions can be depicted which is not permissible in the traditional PCM. The original PCM illustrates both bottom-up (world-driven) and top-down (schema-driven) information processing, but progressing around the perceptual cycle is one-directional through the three phases (i.e. world information → triggers schemata → resulting actions → world). However, there are instances when world information leads directly to actions. For example, in Figure 8.6 the briefing from the WO is world information for the other three crew members, and this links directly to the actions of the PF (conducting pre-winching checks).

Modelling information via a network also allows for patterns in the flow of information and differences between the phases of the incident to be uncovered. For example, in the earlier phases of the incident (Figures 8.5 and 8.6), the information nodes are

TABLE 8.2
Frequency Count of 'From–To' Links for Each Phase

			To		
			Schema	Action	World
Phase 1	From	Schema	–	3	1
		Action	3	3	3
		World	2	4	5
Phase 2	From	Schema	–	3	2
		Action	–	–	7
		World	3	6	4
Phase 3	From	Schema	–	4	–
		Action	–	–	3
		World	5	–	–
Phase 4	From	Schema	–	5	1
		Action	–	1	5
		World	7	2	–
Phase 5	From	Schema	2	3	2
		Action	–	–	6
		World	4	5	1
Phase 6	From	Schema	–	6	1
		Action	1	2	2
		World	5	1	–
Total	From	Schema	2	24	7
		Action	4	6	26
		World	26	18	10

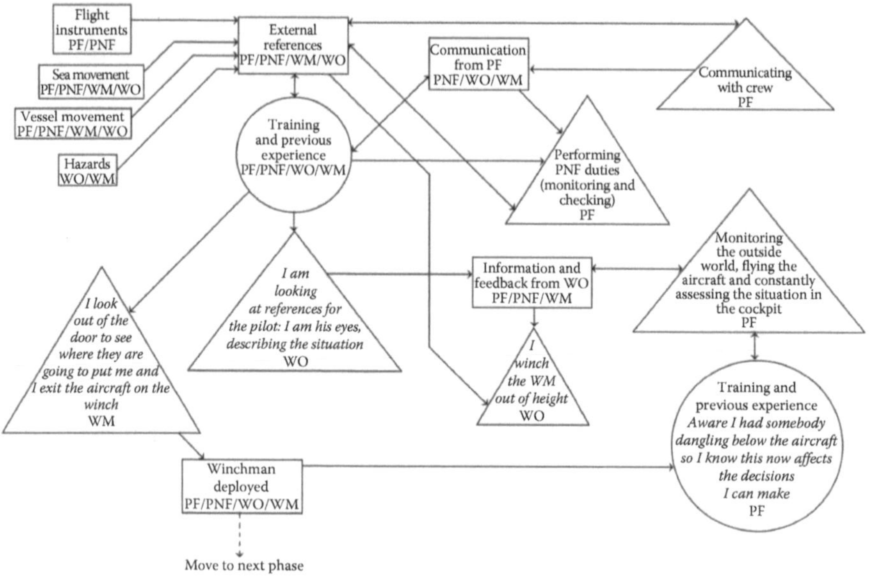

FIGURE 8.6 Network analysis of phase two: live run.

relatively randomly distributed, and there are many interconnections, most of which are bidirectional. However, in the critical incident phase (Figure 8.7), the smoothest flow of information or engagement with the perceptual cycle occurs. One would potentially expect that the critical incident phase would be the most chaotic, but there is a sense from the network that the crew has honed in on only the information they need to deal with the incident, and then each has very specific schemata, resulting in role-relevant duties. This is further supported by data from frequency counts of the node-to-node connections. Table 8.2 shows the node-to-node count for the critical incident phase, and it is clear that these links are only in the traditional direction depicted in the original PCM, that is, schema to action, action to world and world to schema. In contrast, the dummy run phase has a larger spread of connections, both feed forward (traditional) and feedback (counter-cycle), such that there are world to action, action to schema and schema to world links. This will be further discussed in Section 8.4.3.

One of the most interesting aspects of this critical incident in relation to aeronautical decision-making is the demonstration of the role of schemata in decision-making. The role of past experience is emphasised in the PCM, as Neisser proposed that schemata are the medium in which the past affects the future, that is, information previously acquired will determine what will be sampled next. In this incident, the pilots were exposed to the same world information (amber caution light), but they each held very different schemata for the situation (as described in Section 8.3.2.2). This is exemplified in Figure 8.7. Figure 8.8 illustrates the following phase (immediate actions), whereby the action of the PNF telling the PF he thought the event was spurious becomes world information for the PF, but the PF held a schema based on training procedures and previous operational experience that resulted in his actions.

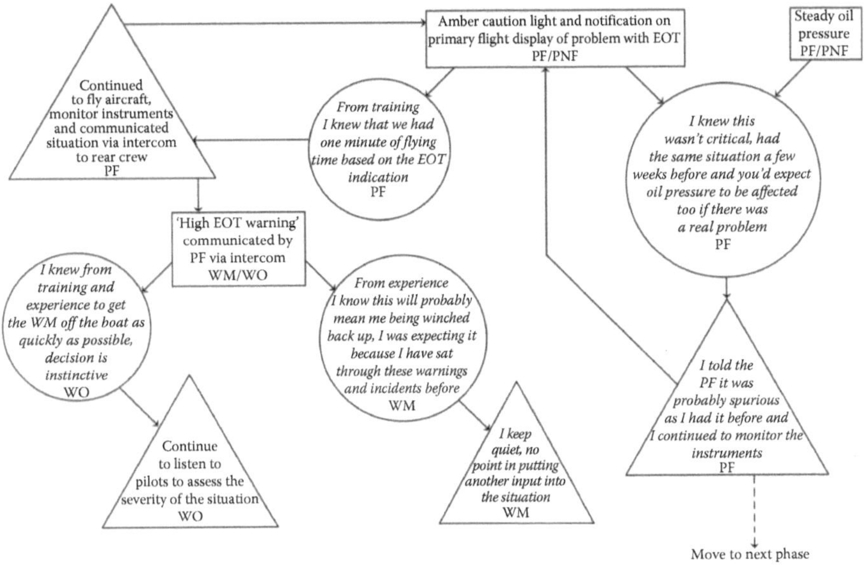

FIGURE 8.7 Network analysis of phase three: critical incident.

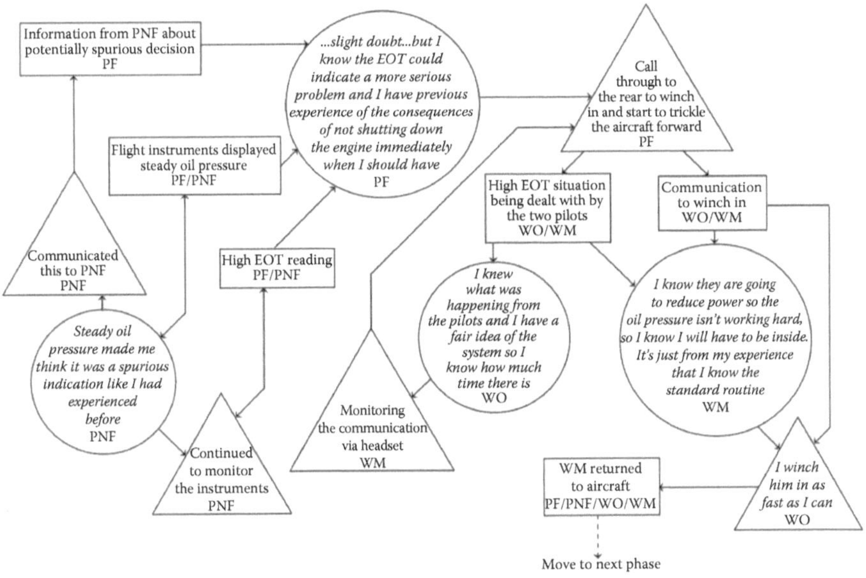

FIGURE 8.8 Network analysis of phase four: immediate actions.

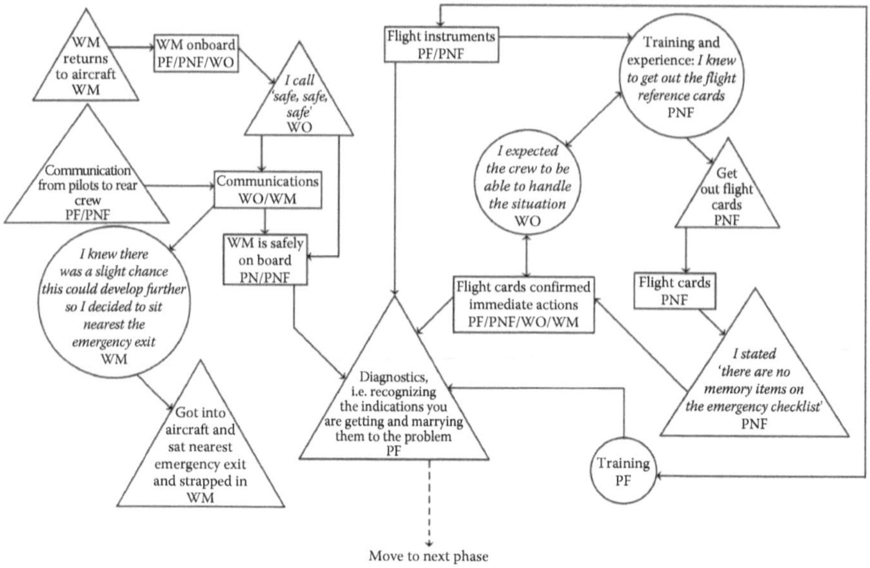

FIGURE 8.9 Network analysis of phase five: diagnostics.

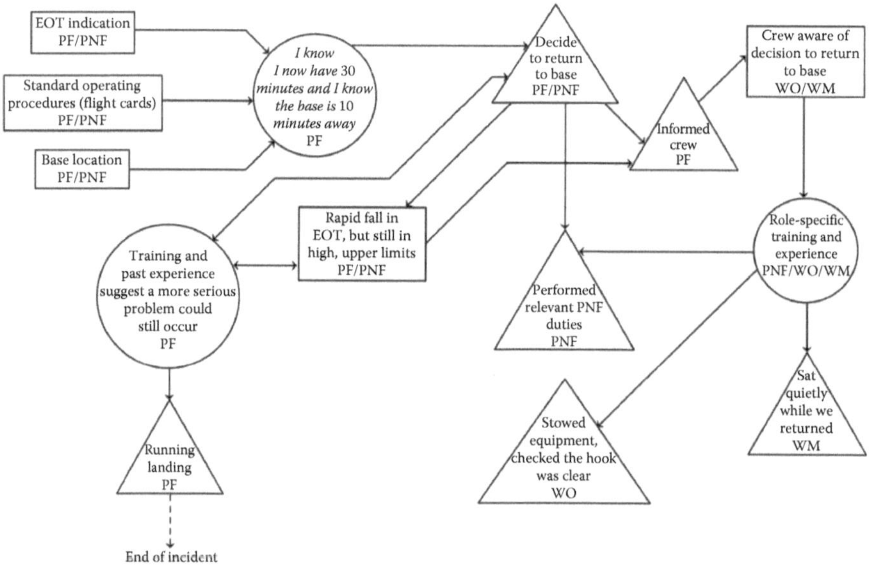

FIGURE 8.10 Network analysis of phase six: return to base.

There is no suggestion here that had the PNF been in command of the aircraft he would have acted counter to standard operating procedures (SOPs) for dealing with such an incident. But it is clear to see how different schemata for the same situation have the potential to be very influential in decision-making, and they may result in instances of inappropriate decision-making. Figures 8.9 and 8.10 illustrate the final phases of the incident, diagnostics and return to base, with the associated frequency counts provided in Table 8.2.

8.4 DISCUSSION

8.4.1 Perceptual Cycle versus Network Representation

The PCM has traditionally been a model of individual cognition, but it recognises that cognition is extended beyond the individual. Distributed cognition is characterised by multiple individuals and technological agents working together to achieve common goals (Hutchins, 1995b). It is generally too simplistic to just consider cognition from the perspective of the individual and often inappropriate to do so in team-based systems. The PCM has been previously applied to complex socio-technical systems that are characterised by the operation of teams, although no attempts have been made to assess whether the PCM is a suitable foundation on which to model a team. The purpose of this chapter was to explore whether the PCM, as an individual-level representation of decision-making, could be applied to a team-level representation. To do this, the CDM was employed to collect decision-making data from four team members of a SAR helicopter team. The three elements of the PCM – schema, action and world – formed the coding scheme in order to qualitatively analyse the data.

In the first analysis (Section 8.3.2), it was demonstrated that the PCM could be used to categorise the decision-making data of individuals within a team, using a method that has previously been applied to understand aeronautical decision-making (Chapter 4). Utilising the perceptual cycle representation has associated benefits; firstly, the cyclical nature of interaction is demonstrated. The individual elements of the PCM and the acknowledgement of both feedback and feed-forward loops set it apart from other information processing and decision-making models. It provides an explicit description of the three elements of information processing (i.e. schema, action and world) and allows data to be structured to demonstrate engagement in the perceptual cycle at a given time. However, when applied to team data, the PCM representation did not demonstrate the interconnectivity that exists between team members working together on a common activity, suggesting that the structure is too rigid to represent the dynamic nature of teamwork. Within any episode of behaviour, an individual evolves through a number of perceptual cycles, each with a cycle of sampling, modifying and directing, but this is not clearly depicted in Figures 8.2 and 8.3.

As a result of the limitations with the traditional perceptual cycle representation, a network-based approach was explored. Networks were chosen because they have demonstrated utility in other domains at representing complex interactions through their depiction of interconnected elements (Walker et al., 2010; Stanton, 2013).

The use of networks in this analysis was novel in that the nodes were coded as the three components of the perceptual cycle (schema, action and world) and directional arrows were used to demonstrate information flow. This allowed the networks to account for the benefits of the perceptual cycle representation, that is, cyclical flow of information and the individual components of schema, action and world, whilst increasing the explanatory nature of the representation by accounting for interconnectivity. This provides a way of extending beyond the individual to team cognitive functions. Neisser (1976) presented the view that perception is an active, rather than passive, process. Activated schemata direct where people look, what people attend to and what people expect to see. This exploration leads to adaption in the environment by the perceiver, which guides further exploration in a cyclical process. Furthermore, Neisser (1976) argued that schema interacts with the temporal nature of events by linking the past to the future via the anticipation of what will happen next determining what we do, and in our understanding of events that occur through that interaction between pre-existing schemata and world information. The unrestricted nature of a network-based analysis allows time to be more accurately represented. The structured nature of the traditional PCM has been previously criticised for not adequately depicting the progression of the perceptual cycle over time, and previous analyses have used sequential numbering to structure the decision-making process (e.g. Chapter 3). However, this approach does not demonstrate the seamless flow of information processing that occurs. The network representation complements the cyclical nature of the original model and allows the progression through the perceptual cycle over the phases of the incident to be represented.

The analysis presented here also offers support to the nature of DSA. As previously outlined, in the DSA approach team schema are considered compatible, rather than shared, because every individual in a team holds their own mental representation of the situation (Stanton et al., 2009a). These individual schemata can combine to form compatible team schemata. Similarly, individuals can be exposed to different world information but through the process of communication the team possess team world information. Also, actions undertaken within a team are coordinated to the extent that team members are working towards a common goal, but individuals within a team will still perform actions relevant to their role and specific tasks. Artman and Garbis (1998) argued that communication and coordination are the most important constructs to study in order to understand successful team performance within a distributed cognition framework. The network perceptual cycle representation allows these interdependencies and independencies to be captured by the role coding of the nodes. Independent nodes (i.e. an individual schema or action or world information relevant to only one person) are coded by just the crew member the node relates to. However, the interdependencies are represented via multiple crew members' coding of the nodes. This is a further advantage of using a network-based approach over the traditional PCM representation.

8.4.2 Modelling Distributed Cognition

Hutchins (2000) argued that the distributed cognition approach does not expect all cognitive events to be encompassed by an individual. Instead, cognitive processes may be distributed in three ways: (1) socially across team members, (2) distributed across internal and external information and (3) distributed through time such that products of earlier events can transform the nature of later events. The perceptual cycle networks presented here account for the three elements of distributed cognition. Firstly, the networks show how information is distributed across different team members and the interconnected nature of this distribution; for example, the actions of one team member can become world information for another team member. Secondly, the use of the PCM elements (schema, action and world) to define the nodes of the networks allows the internal (schema) and external (world) information to be depicted. Finally, the network approach allows the evolution through the perceptual cycle over time to be represented, and the role of schemata demonstrates how past experiences and events can influence actions in the present. This was evident in the different representations held by the PF and PNF about the cause of the critical incident.

In recent years, network modelling has been a common way to represent distributed cognition (Walker et al., 2010; Stanton, 2014). Baber et al. (2013) used networks in the form of social network analysis and agent-based modelling to explore inter-agency communication in a generic SAR scenario. On the basis of a generic SAR scenario, it was shown to be possible to describe the manner in which information flowed between the agents involved in the response. The requirements for the information flow were captured from SOPs that applied to the scenario. However, the use of SOPs alone meant that internal constructs such as schemata of different agents in the scenario were not captured as they were when the CDM data collection technique was utilised in this study. Conversely, by just using the CDMs, it is possible that external information was missed if it was not recalled. Best practice would be to combine the SOP documents and interviews with operators.

Baber et al. (2013) argued that a network analysis approach focuses on explicit external connectivity between agents and makes no assumption about the internal cognitive state of the agents. As such, the connections between agents show the flow of information that the system maintains but says nothing about how the information might be interpreted or used to change the understanding of the individual agents. The analysis presented here has demonstrated that through data collected from CDM interviews, an understanding of each operator's perspective can be represented. Walker et al. (2010) utilised the CDM interview when studying distributed cognition in air traffic control and argued that the CDM interview gives access to some of the unobservable artefacts and interactions that are not captured through observation or SOP documentation. Furthermore, Salmon et al. (2009) argued that the CDM is a useful method to identify information requirements and how this is distributed around a team.

8.4.3 Types of Information Processing

The node-to-node frequency account provides an insight into the kinds of information processing that is occurring and the nature of distributed cognition. Figure 8.11 models this node-to-node frequency count for the whole incident (numbers indicate frequency count totals; see Table 8.2). The findings support the results of the analysis of the validity of the PCM presented in Chapter 5, that is, the solid line

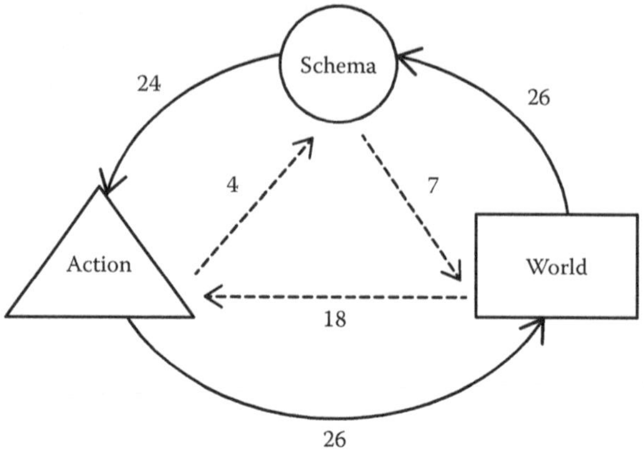

FIGURE 8.11 Model of node-to-node frequency count for the whole incident.

arrows are the interactions following the original PCM flow of information processing (world → schema → action), whereas the dashed arrows represent the counter-cycle (world → action → schema). The majority of node-to-node links fits into the original PCM, with information processing occurring in a feed-forward manner, which was also found in the analysis in Chapter 5. This provides further support for the use of the PCM as a framework for understanding, modelling and explaining decision-making processes, particularly as the data in this chapter were generated from team data and therefore provide preliminary support that the PCM is also suitable for modelling team processes.

As described in Chapter 5, Rasmussen's (1983) Skill-rule-knowledge (SRK) taxonomy can help explain the counter-cycle. In summary, SRK describes three different levels of cognitive control that can be exercised by an individual over their actions. Skill-based behaviour (SBB) refers to the smooth execution of highly practiced largely physical actions where there is virtually no conscious monitoring. Skill-based behaviour develops through extensive practice in which operators develop cued response patterns suited for specific situations. In the network representation of the PCM, SBB manifests as direct links from world information to actions. There are 18 instances in the data (Table 8.2) where world information links directly to action. The specific instances of these interactions are depicted throughout Figures 8.5 through 8.10. For example, the WM re-entering the aircraft (phase 5, Figure 8.8) initiates a skill-based response in the WO (calling *safe, safe, safe*). The highly routine action of calling safe when the WM re-enters the aircraft is executed largely without conscious thought.

A further category of information processing is rule-based behaviour (RBB), and this is characterised by the application of rules (i.e. if x happens, then do y). Rule-based behaviour manifests in the original PCM as bottom-up information processing as the need for action is first recognised due to the state of affairs in the world; this is followed by the retrieval of past rules or methods (i.e. rules are learnt through past experience and stored as schemata), and finally, application of the rules occurs (action). Rule-based behaviour is slower and more cognitively demanding than SBB. In the PCM network representation, RBB presents itself in some of the world → schema → action links. For example, in Figure 8.5 (phase 1, dummy run), the world information of external references (e.g. weather conditions, sea movement, location of potential hazards, etc.) causes the WO to implement RBB for risk assessment (*there is a formula that I go through for risk assessments learnt through years of experience*), and this results in his action of assessing the hazards to conduct the briefing. Rule-based behaviour is also evident during the critical incident phase (Figure 8.7). The PF employs the rule that if the EOT is in the red upper limits (determined by the information presented on the cockpit displays), then that means there is 1 minute of flying time remaining (the rule has been developed through training and is stored as a schema), and this results in his action of continuing to monitor the instruments and fly the aircraft and communicate the situation to the crew. As previously described, this phase of the incident represents the smoothest flow of engagement with the traditional perceptual cycle evident in Table 8.2.

Knowledge-based behaviour (KBB) involves complete conscious activity. This usually occurs in a situation where a beginner is performing a task or when an experienced individual is faced with a completely novel situation. In these situations, the operator has to exert considerable mental effort to assess the situation, resulting in the slowest response times as each action is reviewed and assessed. To generate appropriate responses to novel situations, an operator needs to integrate more general knowledge concerning the behaviour of the system, the characteristics of the environment and the goals to be achieved (Drivalou and Marmaras, 2009). In the PCM, KBB would manifest as the top-down flow of information processing, whereby actions are driven by stored schemata and the resulting change on world information is monitored. KBB requires considerable feedback and this is accounted for in the PCM by the modifying effect that world information can have on the future utilisation of schemata.

8.4.4 Applications, Limitations and Avenues for Future Research

At the beginning of this chapter, it was argued that the PCM offers a more comprehensive account of decision-making than the popular RPD model, owing to its acknowledgement of the cyclical nature of interaction between the mind and the environment. That said, the RPD model serves as the prototypical NDM model and its applicability has been replicated in a variety of domains (Lipshitz et al., 2001). The PCM has the potential to complement the RPD model and to enhance its explanatory power by the inclusion of the role of the external environment. The inclusion of the elements of the PCM, particularly in a network-based approach, also has the potential to enhance Klein's (1999) conception of team decision-making, which only focuses on the team mind and does not acknowledge the role of the external environment, whether that be in the form of physical artefacts available or the influence of other team members on each other. Kirlik and Strauss (2006) argued that the NDM domain achieves an understanding of how people make decisions by placing the human and their psychological processes at the centre of investigation. It is argued that the human and their environment should be the centre of the investigation as their intrinsic interaction with each other means that neither can be studied in isolation. Original ethnographic techniques (e.g. Hutchins, 1995a) were labour intensive. The work presented here provides a relatively quick insight into the distributed nature of team work by showing the interactions that occur between team members and the utilisation of internal and external information.

Utilising a distributed network approach to understand team decision-making has potential utility for collaborative design (Stanton et al., 2009). It is clear from the SAR critical incident that each team member had role-specific tasks, knowledge and experience. Interface displays could be enhanced by presenting only the information that is required by each team member, and potentially this could expand into functional displays that present information relevant to the phase of critical incident. Furthermore, the network-based approach presented in this chapter could be utilised by designers to ensure they have a grasp on what information is utilised by each team member and how this information is used. This is achieved through the mapping of different information elements, both internal and external to the team members, within the networks. The interactions between the information elements of different team members also have the potential to aid designers in creating functional displays that support how different pieces of information are linked and utilised by different team members which would translate into how these are grouped and presented on an interface. Potential applications also exist in the form of enhancing decision-centred training for teams. For example, having an understanding of the way successful teams engage in the perceptual cycle will allow training to be targeted at less successful teams in terms of the processes that can be employed to emulate successful teams.

In the network analysis undertaken by Baber et al. (2013), the wider inter-agency team was considered through social network representations using weighted arrows to demonstrate the strength of connections between various agents. For example, the helicopter captain was linked most strongly to the maritime rescue coordination centres, then the WM and RAF station operations, then the hospital and, finally, the lifeboat cox. The data presented in this chapter have just explored a team perceptual cycle for the immediate aircraft crew. Future endeavours will need to consider how the wider agencies external to the aircraft crew impact on a team perceptual cycle. The agents that the helicopter captain is connected to provide world information, and this has the potential to modify any schemata held for the situation and any subsequent actions. Weighted connections can also be explored in future analyses to enhance the explanatory power of a prototypical network model. It will also be interesting to explore different types of teams such as those with a strong authority gradient or operate as ambiguous teams. Levi and Slem (1995) have acknowledged that teams operate under a range of structures, from leader-controlled decision-making to consensus decision-making. Although a descriptive level of analysis is offered from the approach presented here, Walker et al. (2010) argued that network modelling can offer predictive insights because networks can be subjected to known changes, for example information structure, and outputs can be derived as to the effects of these under various performance contexts.

8.5 CONCLUSIONS

The use of the network method has the potential to be applicable to any domain, and the graphical data outputs are easily interpreted (Walker et al., 2010). In summary, the following information is captured in the network analysis, which is not possible in the traditional PCM representation:

1. The interconnectivity between the elements of the PCM and different crew members is highlighted via the directional arrows and the crew role allocation to nodes.
2. Automatic actions can be demonstrated via direct connections between world information and actions (bypassing cognitive schema).
3. The amount of nodes relating to each component of the PCM (schema, action or world) provides an indication as to the type of information processing (top-down or bottom-up) occurring at different phases of the incident or between different crew members.
4. Patterns in the flow and distribution of data can be explored; for example, there are clear differences between the single-cycle representation during the critical incident (Figure 8.7) and the multicycle representations in the phases (Figures 8.5, 8.6, 8.8 through 8.10).
5. The way the crew members progress through the perceptual cycle over time is more seamlessly represented.
6. Size of networks can enable inferences to be made about crew workload. For example, larger networks suggest that the crew members are engaged with a number of perceptual cycles within one phase.

The PCM demonstrates that teams are not linear information processors, but instead behaviour is a product of the amalgamation of data-driven (world) and knowledge-driven (schemata) information. The PCM has underpinned a variety of research domains which generally focus on issues such as performance or accident analysis in complex socio-technical systems, in which the role of teamwork is critical. However, until now a discussion of teamwork and the PCM has been limited. This chapter presented a case study from the SAR helicopter domain to establish whether the PCM could be used to structure and model team data from CDM interviews. It was found that the PCM coding scheme could be used to analyse team data but was not an appropriate way to structure or model team data. Instead, network perceptual cycle models were created to depict the interconnectivity that is essential to understanding how a team engages in the perceptual cycle. Members of a team share common goals and act interdependently; however, division of labour still occurs and so team members possess independencies in their tasks and knowledge. The results also provide further support for the findings presented in Chapter 5, demonstrating that automatic, skill-based behaviour features heavily in team decision-making which is something that a comprehensive team PCM needs to represent. Chapter 9 builds on this work by continuing to investigate team perceptual cycle interactions. A different data collection method will be used to provide a comparison previously described methods.

9 Exploring Distributed Cognition in Search and Rescue Teams

9.1 INTRODUCTION

Chapter 8 introduced the idea of exploring team perceptual cycle interactions. This was investigated through retrospective interviews conducted via the CDM with a SAR team that dealt with a critical incident. The analysis of this data demonstrated that a network-based analysis approach was useful for exploring the interconnectivity of team perceptual cycle processes. The research presented in this chapter builds on the work undertaken in Chapter 8 by continuing to explore team perceptual cycle interactions and employs a more robust network analysis approach by utilising other stages of the event analysis of systemic teamwork (EAST) methodology. The context of this research remains in SAR teams, but there are two key differences compared to Chapter 8; first, data were collected via concurrent observations and communication recordings rather than retrospective interviews, and, second, the research is exploring perceptual cycle interactions of the team in general, not specifically related to critical decision-making. In doing so, this research demonstrates how the SAR team functions within a distributed perceptual cycle whereby the actions of one team member become world information for another team member.

9.1.1 PCM, Teams and Distributed Cognition

Neisser's (1976) PCM has been comprehensively explained in the preceding chapters, and Chapter 8 summarises the role of the PCM in relation to team decision-making (TDM). Despite the widespread application of the PCM framework in Ergonomics research (see Chapter 2), the framework has not been explicitly applied to team decision-making, which is surprising given the prevalence of teamwork in safety critical systems. The PCM presents a distributed view of information processing, acknowledging the cyclical interaction that exists between a person and their environment, rather than considering the two parts in insolation. In Chapter 4, an argument was made for the importance of using the PCM to gain a process understanding rather than focusing on outputs of performance. Paris et al. (2000b) argued that the same holds when considering teams; understanding how the task is accomplished can help diagnose performance problems that may inhibit the desired outcomes. A recent application of the distributed cognition perspective has been to the study of situation awareness (SA). Based on the view that SA resides in neither the individual nor the world, but through the interaction of the two, Stanton et al.'s (2006, 2009a) theory of DSA is founded upon the theoretical concepts of Schema Theory, the PCM and the distributed cognition approach. As such, DSA argues that awareness is distributed across human and technological agents involved in collaborative activity (Salmon et al., 2008a).

Stanton et al. (2006) argued that the PCM offers a more comprehensive description of how SA is developed and maintained when compared to Endsley's (2000) three-level model of SA (see Chapter 2). Schemata are central to this view of SA; they facilitate the anticipation of situational events, direct attention to cues in the environment and direct courses of action. Checks are made to establish whether the resulting environmental experience aligns with anticipations and unexpected events facilitate further search and exploration and modify the existing schema. The central role of schemata has implications for the explanation of team SA, which is concerned with purposeful behaviour, defined by Stanton et al. (2009a, p. 482) as

> ...teams working together on tasks towards some end goal in highly dynamic [and] collaborative environments.

Many attempts have been made to explain team SA in collaborative environments, and in doing so, much has been made about the importance of 'shared' SA (see Endsley, 2000; Endsley and Jones, 2001), that is individual team members share an understanding of the same situation, and in doing so, during team activity, SA overlaps between team members so that each individual is aware of their own role-specific SA elements and those of other team members. However, if SA is viewed through the lens of the perceptual cycle, then the concept of shared SA is problematic due to the individual and unique nature of schemata. Even if individual team members are exposed to the same world information, they will be engaged in their own perceptual cycle driven by their personal schemata and resulting in their role-specific actions (Stanton et al., 2015). This was highlighted in Chapter 8, where members of the SAR team demonstrated independencies in their tasks and knowledge, but acted interdependently for the achievement of common team goals. For this reason, Stanton et al. (2009a) argued that SA cannot be shared and instead DSA focuses on the concept of 'compatible' SA between team members, in which there are different, but compatible, SA requirements and purposes. Sorensen and Stanton (2011) experimentally verified this by demonstrating that measures of compatible SA revealed differences between two experimental conditions, whereas measures of shared SA did not reveal any differences, suggesting that SA is distributed between team members, rather than shared.

In recent years, DSA has been increasingly used to study SA in the context of teams and collaborative environments in a variety of domains, including command and control (Stanton et al., 2009a,b; Sorensen and Stanton, 2011, 2013; Rafferty et al., 2013), energy distribution (Salmon et al., 2008a), naval warfare (Stanton et al., 2006) and driving and road user behaviour (Salmon et al., 2014a,b). Stanton et al. (2009a) made reference to a team perceptual cycle during their observations in naval control room simulators and argued that the perceptual cycle approach is the most appropriate for considering team SA (Stanton et al., 2009b).

For team effectiveness to be properly optimised, all of the components that might affect team performance must be understood (Salas, 2005). Team perceptual cycle interactions have not been explicitly explored however. This chapter seeks to build on the initial research presented in Chapter 8 to determine what insights can be gained about team processes when they are investigated from the perspective of the PCM.

9.1.2 Using EAST to Analyse Distributed Cognition

Distributed cognition is described in Chapters 1 and 7. One of the key concerns with taking a distributed perspective is the complexity in measuring and modelling it (Stanton et al., 2009; Walker et al., 2010). Rybing et al. (2015) argued that team methods generally focus on effectiveness and outcome measures, rather than the processes involved, the latter being the important part in understanding team distributed cognition. A comprehensive review of measuring team cognition is provided by Cooke et al. (2004); however, the methods that are discussed are only relevant to simulated task environments and reduce many team parameters to measurable metrics.

Few methods operationalise theoretical concepts from the distributed cognition literature (Rybing et al., 2015). One of the few examples is distributed cognition for teamwork (DiCoT; Blandford and Furniss, 2006). This is a socio-technical framework that emphasises how physical artefacts, such as workplace arrangements, actors and artefacts, facilitate collaboration and cognition. DiCoT has predominantly been applied in a clinical setting (Rybing et al., 2015). Networks have demonstrated previous utility at measuring distributed cognition because they are able to represent complex interactions through their depiction of interconnected elements (Walker et al., 2010; Rafferty et al., 2013; Stanton, 2014). In support of this, the preliminary research undertaken in Chapter 7 demonstrated that a network-based approach was the most appropriate for representing team perceptual cycle interactions.

The EAST (Stanton et al., 2008) methodology is a contemporary approach for analysing and modelling distributed cognition. EAST is underpinned by the notion that complex collaborative systems can be meaningfully understood through a network-of-network approach. Specifically, three networks are included:

1. Task networks describe the relationship between tasks and their sequences.
2. Social networks analyse the communication structure (relationships) and the communications (activity) that occur between the different agents (both human and non-human) in a team.
3. Information networks describe the information that different agents use and communicate during task performance.

Each network analysis process has been developed and applied independently, but the unique advantage of EAST lies in bringing the three networks together in the same analysis framework (Stanton, 2014). Since its development, the method has been employed in a number of areas, including aviation accident analysis

(Griffin et al., 2010), air traffic control (ATC) (Walker et al., 2010), naval operations (Stanton, 2014) and military command and control (Walker et al., 2006). As described in Chapter 8, the original version of EAST required data collection from a number of sources, but Stanton (2014) utilised a shortened version in which networks are developed directly from raw data. This approach was undertaken in the previous chapter and will also be employed here; the analysis is also extended beyond Chapter 8 by the inclusion of all three network analyses that encompass EAST, whereas Chapter 8 only considered the information network.

The analysis in this chapter is further enhanced through the application of network analysis metrics via the Agna™ software (version 2.1.1). As described in Chapter 6, the metrics were traditionally developed for social network analysis, but in recent years have become increasingly popular as a way to gain deeper, quantitative, insights on qualitatively derived networks. Table 9.1 describes the available metrics, domain of application and additional references for interested readers. Salmon et al. (2014a) acknowledged that the metrics are not being applied to the network types they were developed for; however, they argued that they offer an appropriate and repeatable way to make inferences about networks. Stanton (2014) proposed that metrics should be selected based on the evaluation being performed and that not all metrics are relevant to all research questions. In this chapter, the metrics of density, cohesion and diameter were calculated for the whole network and for individual nodes the metrics of emission, reception and sociometric status were calculated.

9.1.3 Teamwork and Communication in Cockpit Crews

Teams consist of multiple individuals working within a common organisational perspective; it is rare that in complex socio-technical systems an operator will ever act entirely alone. Team performance is paramount to system safety, and the resulting output is usually greater than the sum of the individual parts (Paris et al., 2000). In aviation, accident and incident reports repeatedly highlight the relationship between good teamwork and safety (Annett and Stanton, 2000; Grote et al., 2010; Bourgeon et al., 2013). Annett and Stanton (2000) highlighted communication and coordination as crucial processes for accomplishing team tasks. In support of this, Sexton and Helmreich (2000) reported on a study that suggested that pilot error was more likely to reflect failures in team communication and coordination than deficiencies in technical ability. As such, extensive research efforts have been devoted to fostering successful teamwork interactions, but Artman and Garbis (1998) argued that little is known about the actual mechanisms underlying the situated cooperative process that enables successful team performance. These processes, however, are not as readily quantifiable as team inputs and outputs and therefore present a harder research challenge (Paris et al., 2000).

There are some studies that have sought to address this. For example, Grote et al. (2010) identified three characteristics of advanced and effective team interaction as being implicit coordination (this occurs when the team have compatible awareness of task requirements, this allows team members to predict each other's intentions and act on them without explicitly communicating strategies and plans), heedful

TABLE 9.1
Metric Descriptions

Metric	Description	Domain of Application (References)
	Analysing the Whole Network	
Density	Network density describes the comparison between the number of interconnections that are possible (i.e. every node connected to every other node) and those that are actually observed (can be represented as a fraction of the total possible). It explores the interconnectivity of the network as a whole and ranges from 0 (no connection between concepts) to 1 (every concept connected to every other concept). High levels of interconnectivity suggest a richer set of semantic links and a well-integrated set of concepts. A denser network is likely to have better connected concepts and shorter average path lengths.	Military command and control (Walker et al., 2009a,b; Rafferty et al., 2013; Sorensen and Stanton, 2013) Naval operations (Stanton, 2014) Driving (Walker et al., 2011; Salmon et al., 2014; Banks and Stanton, 2015) Search and rescue (Baber et al., 2013) Rail (Walker et al., 2006)
Cohesion	Network cohesion describes the number of reciprocal connections in the network divided by the maximum number of possible connections.	Naval operations (Stanton, 2014) Driving (Banks and Stanton, 2015)
Diameter	Network diameter refers to the largest geodesic (i.e. the shortest possible line between two points on a sphere) distance within a network. This can be thought of as another metric of the networks size (eccentricity). The bigger the diameter, the more concepts within the network that exist on a particular route through it. Generally, more dense networks will have a smaller diameter (because routes across network are shorter and more direct), whilst a less dense network will have a larger diameter.	Military (Sorensen and Stanton, 2013) Naval operations (Stanton, 2014) Driving (Walker et al., 2011; Banks and Stanton, 2015)

(Continued)

TABLE 9.1 (*Continued*)
Metric Descriptions

Metric	Description	Domain of Application (References)
	Analysing Individual Nodes	
Sociometric status	Sociometric status refers to the relative importance of a node and relates to the number of links received and emitted relative to the number of nodes in the network and therefore relates to connectedness of a particular information element. It can also be considered in terms of 'busyness', i.e. how 'busy' the concept is relative to the total number of concepts within the network under analysis. It is a useful metric to use to make comparisons in the data: Compare mean change of sociometric status under different conditions (i.e. trials or phases) to see if a node is more or less prominent in the different conditions The mean value can be used as the cut-off point and concepts that fall above the mean can be considered key concepts in a network	Military command and control (Walker et al., 2009a,b; Sorensen and Stanton, 2013) Naval operations (Stanton, 2014) Energy distribution (Salmon et al., 2008) Driving (Salmon et al., 2014; Banks and Stanton, 2015) Emergency services incidents (Houghton et al., 2006) Rail (Walker et al., 2006)
Emission	Number of links exiting a node	Naval operations (Stanton, 2014)
Reception	Number of links entering a node	Naval operations (Stanton, 2014)
Eccentricity	The largest number of 'hops' to get from one side of the network to the other, i.e. distance from a node to the furthest node from it	Naval operations (Stanton, 2014)

(Continued)

TABLE 9.1 (*Continued*)
Metric Descriptions

Metric	Description	Domain of Application (References)
	Analysing the Whole Network	
Centrality	Node centrality determines the central or key nodes in a network; it refers to closeness of different information elements, a central node is one that is relatively close to all other nodes. Concepts with high centrality have on average a shorter distance (measured in edges) to other concepts and are likely to be well clustered and near the centre of the network.	Military command and control (Walker et al., 2009b; Rafferty et al., 2013) Naval operations (Stanton, 2014) Energy distribution (Salmon et al., 2008) Driving (Walker et al., 2011) Emergency services incidents (Houghton et al., 2006) Rail (Walker et al., 2006) Aviation (Walker et al., 2010)
Closeness	Closeness is the inverse of the sum of the shortest distances between each node and every other node in the network. It reflects the ability to access information through the grapevine of network members, i.e. how fast information can be spread from the node to all others.	Naval operations (Stanton, 2014)
Farness	Farness is the index of centrality for each node in the network computed as the sum of each node to all other nodes in the network by the shortest path (inverse of closeness).	Naval operations (Stanton, 2014)
Betweeness	Betweeness is defined by the presence of an agent between two other agents, which may be able to exert power through its role as an information broker, i.e. the number of times the node appears on shortest path between all pairings of other nodes.	Naval operations (Stanton, 2014)

Sources: Houghton, R.T. et al., *Ergonomics*, 49(12–13), 1204, 2006; Stanton, N.A., *Ergonomics*, 57(3), 403, 2014.

interrelating (this concerns deliberate efforts made by team members to consider the effects of their actions in relation to the goals and actions of others and the broader context) and shared leadership (any team member can take on leadership functions). Rafferty et al. (2010) reviewed the teamwork literature and identified key factors which were fused to define five core conceptual categories relevant to teamwork. These were communication, cooperation, coordination, compatible schemata and SA. The authors quantifiably demonstrated that these were the most frequently occurring factors within the teamwork literature. Rafferty et al. (2013) went on to demonstrate that communication was the factor that distinguished between more and less effective teams; however, it was not the amount of communication that was important but the quality and content.

Stanton et al. (2009a) argued that the very nature of team performance is such that different team members have different roles and so need to view and use information differently to other team members. Communication is a key aspect of distributed cognition as it is the transaction that allows information to pass between team members and awareness to develop. Successful team performance relies on appropriate information being communicated to the appropriate team member at the appropriate time (Stanton et al., 2009a; Rafferty et al., 2013). The importance of communication in the cockpit is highlighted by the fact that 70% of the first 28,000 reports made to NASA's Aviation Safety Reporting System were related to communication problems (Sexton and Helmreich, 2000). Furthermore, Jentsch and Bowers (2005) argued that every existing theoretical model of team performance has considered some aspect of team communication as a foundation of team performance. Communication is the means by which crew performance is enacted (Orasanu, 1994); therefore, whilst not a central objective of this research, this chapter will devote some attention to the communication processes that occur within the SAR team in order that a more complete understanding of the situation can be gained.

9.2 METHOD

9.2.1 SAR Environment and Crew Roles

Chapter 8 introduced the SAR environment. Data for this study were collected over a two-week period at a SAR base located in the south of England. The captain (C) is the legal commander of the aircraft and has overall responsibility for the flight, the crew and the aircraft. The co-pilot is second in command, and unless the captain becomes incapacitated during the flight, the roles of captain and co-pilot are fixed. However, control of the aircraft is usually shared equally between the captain and co-pilot, with one pilot designated as pilot flying (PF) and the other as pilot not flying (PNF). The PF is responsible for manually flying the aircraft and controlling the autopilot, whereas the PNF has a monitoring role, cross-checking the actions of the PF and being responsible for communications with external agents such as the ATC. When the co-pilot is the PF, the captain retains overall responsibility for the flight. During SAR training sorties, it is common for the roles of PF and PNF to switch during the flight to give both pilots practice and experience in each role.

In addition to the two pilots, a SAR flight crew also consists of the winch operator (WO) and winch man (WM). The WO is in control of the winch, remaining in the aircraft to ensure that the WM is safely deployed and reaches the target location, such as a cliff site, a person in the water or the deck of a vessel. The WO acts as eyes for the PF because the PF will often not have a direct visual reference with the target or WM, and so the WO is responsible for directing the PF to the target and describing the location and actions of the WM. The WO also has communication responsibilities, primarily with the local coastguard (CG), to keep them abreast of the SAR operations. The job of the WM is to be deployed from the helicopter on a winch line in order to reach the target location and coordinate the rescue. During training exercises, the WM is deployed onto a vessel but will not usually recover a 'casualty' unless it is a pre-arranged exercise. If conditions are not suitable to deploy the WM, then weights can be deployed instead via a hi-line (a weighed rope extension to the winch wire that is lowered to the vessel). SAR training serves as training for both the flight crew and the crews of vessels; it is common for training to be a joint venture between the SAR crew and local lifeboat crews to practise formations between the aircraft and lifeboats whilst deploying the WM (or weights).

A four-person SAR crew will work for a 24-hour shift; during this time, there are 4 hours of training available which cover different areas of the operation. Training is valid for 90 days, but the crews try to renew and refresh every two weeks to remain as current and competent as possible (Bond Aviation Group, 2015). When a new crew arrives, there is a thorough shift handover brief between the crews to discuss any issues arising with the aircraft or anything relevant to the operational context. The crew will discuss the weather and how it might affect flying and decide on the training tasks to cover; this is usually dependent on the training needs of the crew and whether they need to cover anything specific.

9.2.2 Participants

During the two-week period, a total of 12 different crews were observed that consisted of 17 individual crew members (4 captains, 4 co-pilots, 4 WO and 5 WM). All crew members were male and very experienced in terms of role-specific hours: the pilots averaged 4400 flying hours (SD = 3300), and the winch crew averaged 4300 role-specific hours (SD = 2000). Ethical permission to conduct the study was granted by the Research Ethics Committee at the University of Southampton. All crew members were aware of the nature of the study and signed a consent form prior to the observation of their training sorties.

9.2.3 Data Collection and Data Treatment

During the research period, the researcher flew with the SAR crews daily and observed a total of 12 training sorties. The sorties were video recorded by either one of the crew members wearing a helmet-mounted camera or the researcher using a video camera with the microphone attached to their headset (the recording equipment was already certified for use in the aircraft as they were previously used in the filming of television documentaries). Each training sortie had a flight time of between 60 and 90 minutes, resulting in approximately 13 hours of in-flight footage which was transcribed into communication transcripts which detailed

- Which crew member was speaking
- Who they were speaking to
- What was said
- Other non-verbal information (e.g. actions undertaken, equipment that was used and observations made)

An example of a transcript is provided in Table 9.2. It was a condition of being granted access to the SAR site that no interventions were made by the researcher, who only observed and recorded the communications. Crew activity time is strictly monitored to ensure that they are suitably rested in the event that they are called to a live task; as such, interview time was not specifically scheduled with the crew members; however, there were occasions during the post-flight brief sessions where clarification questions could be asked.

During the observation period, 10 different types of training tasks were observed. One training sortie usually consisted of 2–3 exercises, and in total, 26 individual exercises were observed. Table 9.3 details the types of training exercises and number of observations. Each training sortie differed in terms of the specifics of the exercise such as what vessel was winched to or the location flown to; however, there were striking similarities between the sorties because very prescriptive task procedures were followed. As such, the observed training sorties have been amalgamated into one representative case study (see Section 9.3). This is an approach that has been previously applied; for example, Baber et al. (2013) used an illustrative scenario to provide a high-level overview of SAR in order to explore the relationship between agencies and their information requirements (see Chapter 8 for more detail).

9.2.4 Data Analysis

The SAR case study resulted in a transcript containing 1544 text segments (either in the form of a direct communication or the annotation of non-verbal activities). The data were analysed using a combination of the EAST methodology, the PCM/SAW coding schemes (see Chapters 3 and 6, respectively) and a communication taxonomy taken from the literature. The analysis and results are discussed in combination in Section 9.4.

9.2.5 Reliability Assessment

As conducted in previous chapters, the reliability of the coding schemes (SAW taxonomy and communication types) was assessed by calculating inter-rater reliability. A detailed description of reliability and justification for calculating percentage agreement is provided in Chapter 4. In this study, to assess reliability, four additional coders were judged by the standard set by the expert coder in a blind condition, that is raters were unaware of the expert's coding decisions. The additional coders had not participated in reliability studies for this research before.

TABLE 9.2
Example of Transcript from SAR Data Collection (during Winch Formation Exercises with a Lifeboat)

From	To	Communication	*Video/Notes/NVC*
PNF(C)	PF	That's your magenta through 60 there, passing 77, magenta approaching 40, passing 70 feet there	
PF	PNF(C)	See what the power does when we get down	
		High 70s, you happy with that PNF?	
PNF(C)	PF	I am yeah	
PF	PNF(C)	Okay that's great	
WO	PF	40 feet?	
PF	WO	40 feet, affirm	
WO	PF	Roger, clear winch	
PF	WO	Affirm, we got 11 knots ground speed on at the moment	
WO	PF	Excellent, winching outboard, the rib's keeping up with us, clearing the step, it is in our 3 o'clock at 4 units	*WM deployed, WO looking out open door of cockpit*
		Getting WM to height	*WO signalling with hands (for the boat, direction to move)*
		WM at height, the boats closing, one unit in our half-past three	
		And that is beautiful	
PF	WO	That's a good one	*PF looking out of window (down and right) at lifeboat*
WO	PF	That is really good	*WO giving hand signals*
		Okay, he's going to keep one unit off, coming again, so he's coming in again, closing, that is quite nice because he is watching me nicely, half a unit	*Boat is practicing running in to WM*
		Nearly there,	
		Beautiful	
		Moving away, that's good	
		Nicely in our half-past three, closing, closing	
		Are you climbing now?	
PF	WO	Aircraft just ballooned ever so slightly	
WO	PF	Closing half a unit	
		That's good, that's nice position	
		One more and then perhaps they're going to change	
		I hope they're going to change	
		He did say he wanted more than one turn	
PNF(C)	WO	Yeah, he did say – hopefully he does, yeah	

TABLE 9.3
Types of Training Exercises and Number of Observations

Training Task Type	No. Exercises Observed
Winch to vessel	7
Winch to cliffs	1
Winch to vessel attempted but discontinued (dummy only)	3
Hook-hoist change over	1
Lifeboat formations	4
ILS approaches and training for pilots	4
Confined area landing	2
Drum winch search and recover	2
Navigation exercise	1
Familiarisation with hospital helipad	1
10 Total	26

To assess the reliability of the SAW taxonomy, the coders were given a training session to describe the different classification categories. The coders were presented with 150 text segments (representing 10% of the data). The text segments were selected using a random number generator. It was found that this threshold was reached with all of the classification schemes (schema subtypes: 83%; action subtypes: 90%; world subtypes: 80%; and communication types: 93%). From this analysis, it is clear that the three PCM categories met the 80% threshold level of agreement for both inter-rater reliability. The communication-type coding scheme also met the 80% threshold for inter-rater reliability. These reliability scores can increase confidence in the use of the coding scheme.

9.3 CASE STUDY SYNOPSIS

The SAR training case study contains the most frequently observed exercises and only those that were relevant to the whole crew; therefore, the instrument flying exercises have been excluded as these were pilot-specific exercises. The training case study consists of 17 phases, including 4 training exercises (winch to vessel, confined area landing, hook-hoist changeover and lifeboat formations). These phases and exercises were observed across the 12 flights and are representative of a SAR training sortie. This was confirmed by the chief pilot at the SAR base who reviewed the transcript and verified its contents as being a representative example of SAR training (minor amendments were made to the technical terminology where necessary).

> The training sortie commences with a shift hand over brief with the previous crew. During this brief, the crew are made aware of any problems arising with the aircraft and briefly discuss the training plan. The captain completes the manual multi-sector load sheet (MMSL). This includes information about temperature, wind, pressure altitude and associated graphs to compute maximum weight allowances and therefore

performance parameters of the aircraft. A time is decided for when the crew will reconvene for the pre-flight brief. This brief occurs just before walking out to the aircraft. The captain briefs the flight crew about the current weather conditions and the finalised training plan. After walking out to the aircraft, the start-up checks are conducted which includes the aircraft being started, the engine and system checks conducted by the pilots, winch checks conducted by the WO and the PF(C) and the before taxi checks. The aircraft then taxi's and before take-off checks are completed during which ATC at the SAR base are contacted and give clearance to take off. Once the aircraft has taken off, after take-off checks are completed by the pilots and the WO contacts the CG to tell them they are airborne.

In the transition phase, the crew are looking for a vessel to winch to. The vessel has requested a winch exercise with the crew, but it is determined that the conditions are not suitable. The aircraft was flown to a ferry in the vicinity to assess its suitability for winching. The pilots conducted the pre-winching checks and the WO opens the door of the aircraft and holds up a board with the numbers 67 written on it. This is to alert the ferry crew to listen to communications from the aircraft on channel 67. The WO speaks to the vessel and requests permission to conduct a winch exercise and asks the ferry crew to let them know if they plan to change course. The winching brief is then conducted by the WO, during which he describes any hazards to the aircraft and the WM and the plan in case of emergencies. Next, the dummy run involves the WO directing the PF towards the vessel; the PF manoeuvres the aircraft into a winch position and assesses the conditions. It was realised that the vessel was going to change course and so the aircraft backed off and another dummy run was conducted once the course change had taken place. With all of the crew happy, they commence the live run, in which the WO directs the PF(C) towards the aircraft and the WM is winched out and onto the vessel. During this time, the PF(C) cannot see the WM or the point on the vessel he is being winched to and so the WO acts as the eyes for the pilot, directing him to where he needs to be and describing the conditions and activities that are occurring. Once the WM is on deck, the aircraft climbs and the WO deploys the hi-line so that a stretcher can be winched down to the WM. The WM and stretcher are then recovered to the aircraft. The post-winch checks complete this exercise.

After the winch exercise, the PF(C) and PNF switch roles and the crew fly to an area closer to shore to conduct a hook-hoist change over. As a safety feature, the aircraft is fitted with twin hoist installation in case the main hoist becomes unserviceable. The training exercise simulates a malfunction when the WM is deployed on the wire, the serviceable hoist is sent down and the WM swaps them around. At a certain point between the changeover, the WM is on a hoist with no power on it; this is a very small period of time but the most critical and hence it is regularly trained for. After this exercise, the crew transition to the Isle of Wight to conduct a Confined Area Landing (CAL). Once the aircraft has landed before take-off checks are conducted between the two pilots and during take-off, the PNF(C) reads out the height to the PF and the winch crew observe for obstacles and clearances from trees. Once the after take-off checks are conducted, the pilots switch roles again and the aircraft is flown to an area to conduct formations with a local LB crew. The same process of checks is repeated: cruise checks, pre-winch checks and winch brief. The WM is winched to height and the LB runs in towards the WM to practise positioning for the WM to join the boat. The LB then moves away and another crew member takes over to practice positioning the boat. During this time, the aircraft is hovering at 40 feet. Once the exercise is complete, the WM is winched in and the post-winch checks are completed. The SAR crew then return to base, ATC are contacted to ensure they are clear to land and the aircraft lands back at the airfield and is shut down. The post-flight debrief occurs once the crew are back in base. This is led by the captain but involves a discussion between all the crew members about each training exercise, that is what went well and why certain decisions were made.

9.4 RESULTS

The transcript data were analysed using EAST to explore the task, social and information networks that exist; in addition, a communication analysis was also conducted.

9.4.1 Task Network Analysis

Walker et al. (2010) suggested that the hierarchical task analysis (HTA) method is suitable to describe the tasks in a system. However, Stanton (2014) argued that a network representation was more appropriate because it portrays relationships between tasks that are non-sequential. Ultimately, the task of SAR training is a linear process moving through a set of standardised checks and procedures and through various phases of flight, including the training exercises, before returning to base and landing via another set of standardised checks and procedures. There are clear dependencies within this process; for example, take-off cannot occur without completing the preceding checks; however, within this linear process, there are non-sequential tasks where decisions about whether to go ahead or continue with training are constantly being assessed and reassessed. For example, training exercises can be aborted if the conditions are not right to complete them safely. Furthermore, there is constant dialogue between the aircraft and coastguard to keep each other informed of their operational circumstances.

Figure 9.1 depicts the task network for the SAR training case study and was structured around the tasks defined in the case study described in Section 9.3. Once the aircraft is airborne, the tasks involved in the different training exercises are generally repeated, whether the exercise is a winch to a vessel or a confined area landing, the crew will work through the tasks of target identification, contacting the vessel, pre-winch checks, winch brief, dummy run, live run and post-winch checks. The crew are constantly assessing and reassessing the conditions and situation, and if it is not deemed safe enough, the exercise will be aborted (this occurred during the first winch exercise and during the confined area landing). At different points during the sortie, the captain and co-pilot alternate roles between PF and PNF. During the training sortie, there is regular contact between the WO and CG to keep the CG informed about the operational conditions and the training plan. Once the training exercises have been completed, the crew return to base, going through the tasks of approach brief, descent and land; this is followed by the final phases of shut down and post-flight debrief.

The order of events in the case study was used to construct an association matrix to indicate the from/to links between each task. This matrix was used to calculate the metrics for the task network. In relation to the whole task network, the network density was 0.08 (i.e. a low distribution network with very small connections between concepts) and network cohesion was 0.02 (i.e. a very low level of reciprocal

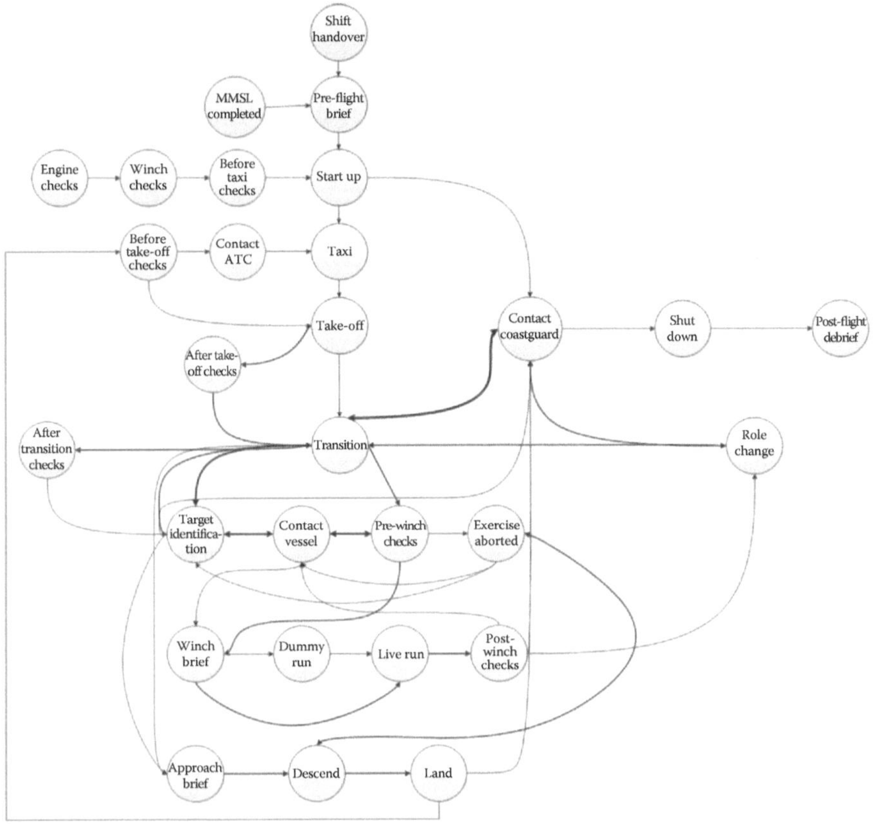

FIGURE 9.1 Task network analysis for SAR training sortie.

links). These two metrics suggest strict dependencies in the task network which is to be expected in the linear and sequential network depicted in Figure 9.1, where one task cannot start without the previous one being completed. The network diameter was 9 (i.e. 9 hops from one side of the network to the other).

The other calculated metrics relate to the individual nodes, and the results are presented in Table 9.4. As described in Table 9.1, it is common practice to use the mean sociometric status as a cut-off point, that is concepts that have a sociometric status above the mean can be considered key concepts in a network (in Chapter 6, the mean plus one standard deviation was used to determine the primary concepts, and the mean was used to determine the secondary concepts). The mean sociometric status was 0.19; 10 of the concepts fell above this threshold (bold in Table 9.4) and can be considered key concepts, including contact CG, transition, after take-off checks, target identification, contact vessel, pre-winch checks, winch brief, live run, role change and descent. The number of links out of, and into, each task node is shown in the emission and reception values. These values were used to depict the strength of connections in the task network. For example, the darker lines in the task network depict a greater number of emission or reception links.

TABLE 9.4
Analysis of Task Network, Concepts Listed in Order of Importance Based on Sociometric Status

Task	Emission	Reception	Sociometric Status
Contact CG	10	9	**0.68**
Transition	9	9	**0.64**
Contact vessel	6	7	**0.46**
Target identification	5	5	**0.36**
Pre-winch checks	5	5	**0.36**
Role change	3	5	**0.29**
After take-off checks	4	2	**0.21**
Winch brief	3	3	**0.21**
Live run	3	3	**0.21**
Descent	3	3	**0.21**
Exercise aborted	3	2	0.18
After transition checks	2	3	0.18
Land	3	2	0.18
Before take-off checks	2	2	0.14
Take-off	2	2	0.14
Post-winch checks	2	2	0.14
Approach brief	2	2	0.14
Before taxi checks	1	2	0.11
MMSL completed	1	1	0.07
Pre-flight brief	1	1	0.07
Start-up	1	1	0.07
Engine checks	1	1	0.07
Winch checks	1	1	0.07
Taxi	1	1	0.07
Contact ATC	1	1	0.07
Dummy run	1	1	0.07
Shut down	1	1	0.07
Shift handover	1	0	0.04
Post-flight debrief	0	1	0.04

Note: Values in bold are above the mean value.

9.4.2 Social Network Analysis

The agent roles in the SAR case study are depicted in Table 9.5. In relation to the pilots, different roles are assumed by the same person, that is the captain can be the PF or the PNF. An association matrix was constructed based on the communication and observational data in the case study. This matrix, presented in Table 9.6, represents the frequency of communication between the different agents in the network, and this is depicted in the social network in Figure 9.2, where the communications are represented by directional arrows and the frequency of communications

TABLE 9.5
Agent Roles in the SAR Case Study

Name	Role
Previous captain	Captain from previous shift
PF(C)	Pilot flying (captain)
PNF	Pilot not flying (co-pilot)
PF	Pilot flying (co-pilot)
PNF(C)	Pilot not flying (Captain)
Pilots	The two pilots in the front of the aircraft
Winch crew	The two winch crew in the back of the aircraft
WO	Winch operator
WM	Winch man
GC	Ground crew
ATC	Air traffic control at SAR base
CG	Coastguard
Vessels	Ships in the sea that the SAR crew make contact with
ARCC	Aeronautical Rescue Co-ordination Centre
System	The systems of the aircraft, communicating via system messages (e.g. a traffic warning)

determine the strength of the connection between agents, that is darker lines represent stronger links based on high frequencies (and vice versa). The communication frequency data were used to compute the social network metrics and are presented in Table 9.7. The network density is 0.2, indicative of a broadly spread network with few links. Cohesion was 0.15, suggesting a low number of reciprocal links, and the network diameter was 4, that is 4 hops from one side of the network to the other. In relation to the individual nodes, sociometric status was calculated to determine the relative importance of each node. Taking the mean value (10.3) as the cut-off, five roles were identified as the most important in the social network: the captain and co-pilot roles of PF and PNF and the WO (bold in Table 9.7).

These five roles are part of the immediate SAR crew, and so metrics were calculated for this network also. Figure 9.3 depicts the social network for the immediate SAR crew and combines the PF and PNF roles (regardless of who flew the role). The network density is 0.8, indicating a tightly connected network. Cohesion was 0.8, suggesting a high number of reciprocal links and network diameter was 2, that is short pathways within the network.

9.4.3 Information Network Analysis

Traditionally, information networks are created by analysing the concepts in communication transcripts. Generally, concepts are identified by keywords (e.g. nouns and verbs), and transition or connection words (e.g. and, or, also) are excluded. In Stanton's (2014) EAST, analysis of how a submarine returns to periscope depth concepts in the transcript was identified and paired with its nearest relation (i.e. concepts in the same

TABLE 9.6
Association Matrix for Social Network Analysis

FROM/TO	Previous Captain	PF(C)	PNF	PF	PNF(C)	Pilots	Winch Crew	WO	WM	System	GC	ATC	CG	Vessels	ARCC
Previous captain	0	1	0	0	0	0	0	0	0	0	0	0	0	0	0
PF(C)	0	0	132	0	0	0	12	90	6	11	3	0	0	0	0
PNF	0	142	0	0	0	0	3	15	0	0	1	0	0	0	0
PF	0	0	0	0	88	0	7	47	1	0	0	0	0	0	0
PNF(C)	0	0	0	101	0	0	2	11	0	0	0	3	1	0	0
Pilots	0	0	0	0	0	0	24	0	0	0	0	0	0	0	0
Winch crew	0	0	0	0	0	0	0	0	0	0	0	0	0	0	0
WO	0	173	17	75	11	3	0	0	7	0	0	0	14	21	1
WM	0	6	0	9	0	2	0	8	0	0	0	0	0	0	0
System	0	0	0	0	0	21	0	0	0	0	0	0	0	0	0
GC	0	1	1	0	0	0	0	0	0	0	0	0	0	0	0
ATC	0	0	2	0	2	0	0	0	0	0	0	0	0	0	0
CG	0	0	0	0	1	1	0	9	0	0	0	0	0	0	0
Vessels	0	0	0	0	0	0	0	18	0	0	0	0	0	0	0
ARCC	0	0	0	0	0	0	0	2	0	0	0	0	0	0	0

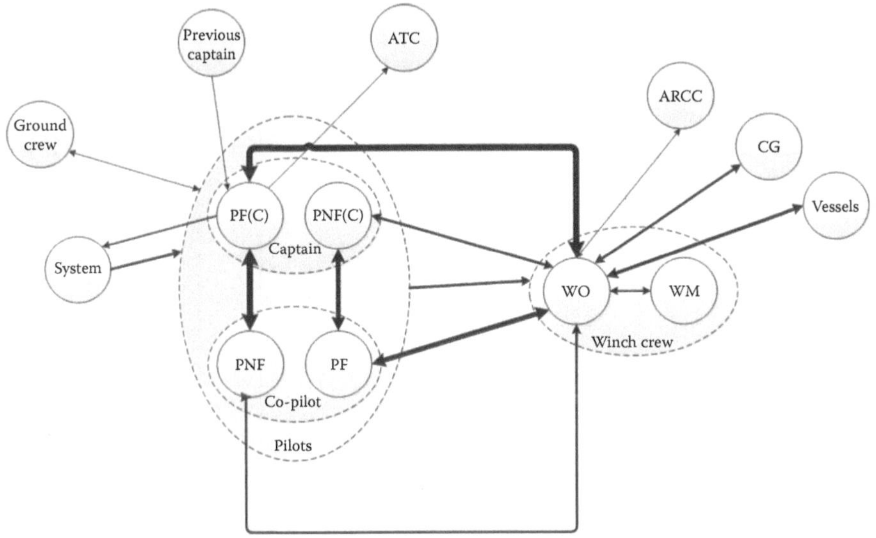

FIGURE 9.2 Social network for the SAR training sortie.

sentence) to build an information network for the submarine control room. A different approach has been taken in the construction of networks presented in this book because the focus of the study is to investigate team interaction in the perceptual cycle; therefore, an additional level of analysis was conducted before the information networks were compiled. The text segments in the SAR communication transcript were qualitatively analysed using the PCM and SAW coding schemes that have been presented in previous chapters. In line with the thematic analysis approach described in Chapter 4, each text segment was firstly coded against the PCM categories of schema, action and world and then against the SAW taxonomy categories, that is subtypes of schema, action and world (see Chapter 6). This approach has the potential to produce more generalisable networks, the benefits of which are discussed in Section 9.5.3.

As described in Chapter 6, the SAW taxonomy was developed from retrospective CDM interviews, whereas the SAR case study was generated from communication transcripts and observational data. Therefore, the first focus of this analysis is to determine the validity of the SAW taxonomy with data collected from different sources. The second aim of this analysis is to explore the team interactions in the perceptual cycle process.

9.4.3.1 Thematic Analysis with the SAW Taxonomy

The SAW taxonomy presented in Chapter 6 was used to thematically analyse the data that had already been coded against the three elements of the PCM (schema, action and world). A key difference that emerged between the development of the SAW taxonomy with retrospective interview data and using it to code communication transcripts was the action subtype: communicate. In the original SAW taxonomy, this was defined as *statements relating to the sharing or exchange of information,*

TABLE 9.7
Analysis of Social Network

	Previous Captain	PF(C)	PNF	PF	PNF(C)	Pilots	Winch Crew	WO	WM	System	Ground Crew	ATC	CG	Vessels	ARCC
Emission	1	254	161	143	118	0	0	322	25	21	2	4	11	18	2
Reception	0	323	152	185	102	27	24	200	14	11	4	3	15	21	1
Sociometric status	0.07	**41.21**	**22.36**	**23.43**	**15.71**	1.93	1.71	**37.29**	2.79	2.29	0.43	0.50	1.86	2.79	0.21

Note: Values in bold are above the mean value.

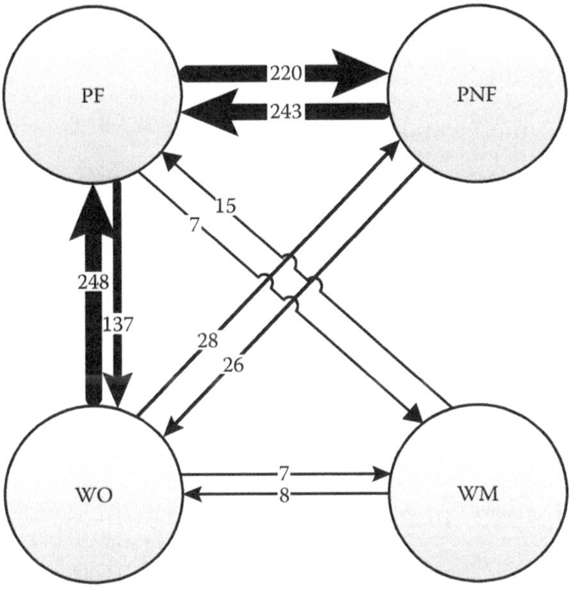

FIGURE 9.3 Social network with frequency of associations between the SAR flight crew.

clearly every transaction in a communication transcript is a communicate action, but a boundary was set to code the content of the communication, unless an explicit communication action was evident, for example contacting the coastguard or asking a question. The communicate category description was redefined as *statements relating to the explicit sharing and exchange of information or expressions used to elicit information*. Similarly, the action category of 'aviate' is constantly happening, but this was only coded if explicit reference was made to it, for example, *alright, flying away then*' (PF). Other changes made to the SAW taxonomy are summarised in Table 9.8 (no edits were made to the schema subtypes). The SWARM prompts introduced in Chapter 7 and provided in Section A.1 reflect these additional categories and can therefore be used for team research too.

9.4.3.2 Perceptual Cycle Interactions: High Level

The first phase of the perceptual cycle network analysis explored the data with the high-level PCM codes of schema, action and world and the roles of the four members of the SAR aircraft crew, in which the captain/co-pilot roles were combined to just the flight role of PF or PNF, that is who was performing that role was not considered because the important element is the flying role and not who is in overall command of the aircraft. For this analysis, the final task of post-flight debrief was removed because pertinent transactions were moved to the relevant section of the transcript and therefore analysis would have occurred twice if the post-flight debrief task was included. The data were coded by PCM category and used to create an association matrix based on the from/to links between the different roles; this is presented in Table 9.9.

TABLE 9.8
Changes Made to the SAW Taxonomy

Original Category	Edit	Justification
Action		
Communicate	Redefined	To make it more suitable to the analysis of concurrent communication data *statements relating to the explicit sharing and exchange of information or expressions used to elicit information.*
System management	Renamed	*System interaction*: interaction implies more of an active influence is exerted whereas management seems more of a passive process.
Checklist items	Added	It was apparent when analysing the concurrent team data (as opposed to retrospective individual data) that many actions were concerned with the completion of checklist items (e.g. before taxi checks, pre-winch checks, post-winch checks, after transition checks) *statements relating to the completion of checklist items.*
Operational	Added	Many actions undertaken were in relation to the operational situation (and were associated with the world category of operational context): *statements relating to actions that are performed for the current or intended operational situation.*
World		
Technological conditions	Renamed	*Technological status*: The original SAW taxonomy was developed from critical incident interviews, whereas the SAR transcripts were collected under normal conditions. The term status is more suitable for normal and non-normal situations. The definition of this category has not changed.
Aircraft status	Redefined	The original definition for this category was: *statements relating to the current status of the aircrafts integrity or performance*. This still holds true but needs to be more generalised for normal conditions where status might refer to whether the door is open or closed or how much fuel there is: *statements relating to the aircrafts integrity, performance and the general state of affairs of the aircraft at a particular time.*
Crew status	Added	The team data produced instances where world information was gained about the status of the crew, for example the WM being ready to be winched out. This category was defined as *statements relating to the readiness and preparedness of the crew or the current status and activities of a particular crew member.*

The data in Table 9.9 were used to compute the metric values for the high-level information network (see Table 9.10). The network density is 0.6, a medium distribution density suggesting that there are moderate connections between the concepts. Cohesion was 0.5 which is the highest value of all the networks and is indicative of a higher number of reciprocal links between the concepts. Network diameter was 3, that is 3 hops from one side of the network to the other. In relation to the individual

TABLE 9.9
Association Matrix for PCM Network Analysis

From/To	PF_S	PF_A	PF_W	PNF_S	PNF_A	PNF_W	WO_S	WO_A	WO_W	WM_S	WM_A	WM_W
PF_Schema	0	14	3	0	7	0	0	3	2	0	0	0
PF_Action	2	0	31	2	82	63	0	81	63	1	6	4
PF_World	18	69	0	1	63	15	0	86	8	0	1	0
PNF_Schema	1	1	1	0	6	1	1	1	0	0	0	0
PNF_Action	2	88	106	1	0	21	0	35	7	0	2	0
PNF_World	0	24	10	8	36	0	0	13	3	0	2	0
WO_Schema	0	1	0	0	0	0	0	2	0	0	0	0
WO_Action	4	87	101	1	38	10	1	0	42	0	14	0
WO_World	1	31	12	0	18	4	1	60	0	0	2	0
WM_Schema	0	0	0	0	1	0	0	1	0	0	0	0
WM_Action	0	4	2	0	4	0	0	6	10	0	0	1
WM_World	0	0	0	0	0	0	0	1	0	1	0	0

TABLE 9.10
Analysis of PCM Network

	PF_ Schema	PF_ Action	PF_ World	PNF_ Schema	PNF_ Action	PNF_ World	WO_ Schema	WO_ Action	WO_ World	WM_ Schema	WM_ Action	WM_ World
Emission	29	335	261	12	262	96	3	298	129	2	27	2
Reception	28	319	266	13	255	114	3	289	135	2	27	5
Sociometric status	5.18	**59.45**	**47.91**	2.27	**47.00**	19.09	0.55	**53.36**	**24.00**	0.36	4.91	0.64

Note: Values on bold are above the mean value.

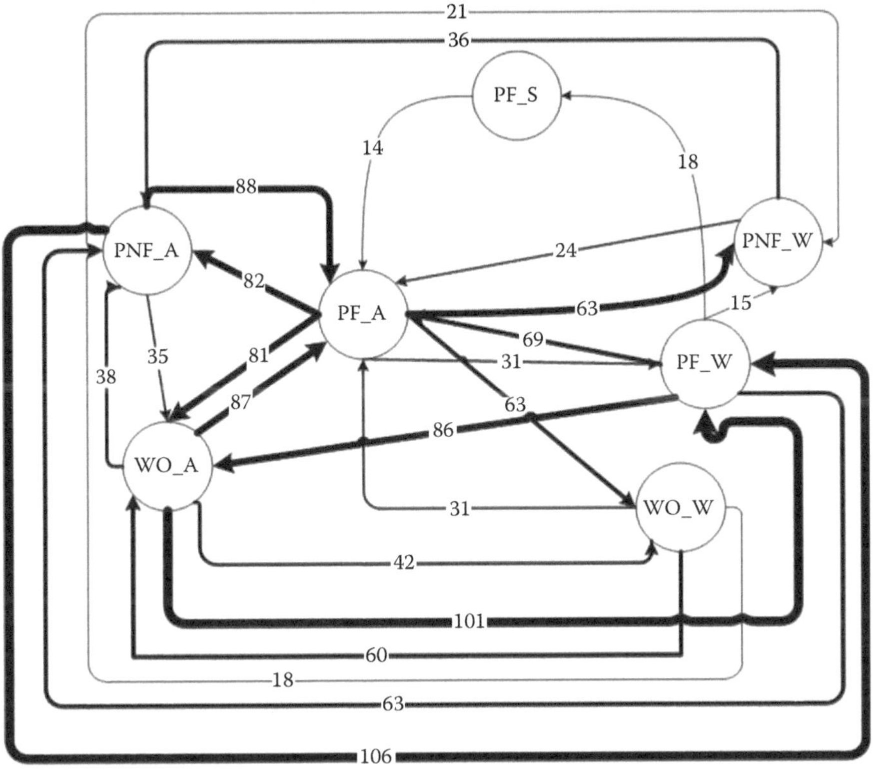

FIGURE 9.4 Network for high-level PCM concepts and role (numbers denote frequency of communications).

nodes, sociometric status was calculated to determine the relative importance of each node. Taking the mean value (22) as the cut-off, five concepts were identified as the most important in the social network: PF_action, PF_world, PNF_action, WO_action and WO_world (bold in Table 9.10). The frequency count values in the association matrix were used to create the information network depicted in Figure 9.4 (note that for clarity only links that represent 1% or more of the data are included).

9.4.3.3 Perceptual Cycle Interactions: SAW Taxonomy

The data were also analysed with the SAW taxonomy, an association matrix of from/to links was compiled for the SAW taxonomy categories; this analysis did not include team role. The association matrix was used to compute the metric values. There were 31 nodes in the network, the network density is 0.4, suggesting there are moderate connections between the concepts. Cohesion is 0.2, indicating a low number of reciprocal links which reflects the spread of data throughout the network. Network diameter was 3, that is 3 hops from one side of the network to the other. In relation to the individual nodes, sociometric status was calculated to determine the relative importance of each node. To align with the SAW analysis presented in Chapter 6, primary concepts were identified as those with a sociometric status value greater than the mean plus one standard deviation (7 concepts), secondary concepts were identified as those with a sociometric status value greater than the mean (7 concepts) and tertiary concepts were everything else (14 concepts). Three concepts were not coded in the data set. The results of this sociometric status analysis are presented in Table 9.11. The data in the association matrix for the primary and secondary

TABLE 9.11
SAW Concepts Ordered by Sociometric Status Value (Highest to Lowest)

SAW Concept	Sociometric Status Value
A_Enviro.Mon	9.53
A_Checklist	8.17
A_Aviate	7.90
W_Location	7.33
A_Syst.Mon	6.67
A_Operational	6.57
A_Navigate	6.23
W_Op.Cont	**5.47**
A_Syst.Int	5.40
A_Communicate	5.33
W_Tech.status	4.67
A_Decision	4.53
A_Sit.Ass	3.83
W_Display Inds.	3.10
W_Comm.Info	**2.97**
W_Aircraft status	2.67
A_SOP	2.50
W_Nat.Envio conds	2.13
W_Artefacts	1.70
W_Crew status	1.20
W_Physical cues	1.17
S_Declarative	1.07
S_DPE	0.73
S_TPE	0.47
S_Analogical	0.47
W_Absent info.	0.20
S_VPE	0.13
S_Insufficient	0.13
A_Con.Diag	0.00
A_Non-Action	0.00
W_Problem severity	0.00

Note: Bold font denotes the cut-off between primary, secondary and tertiary concepts.

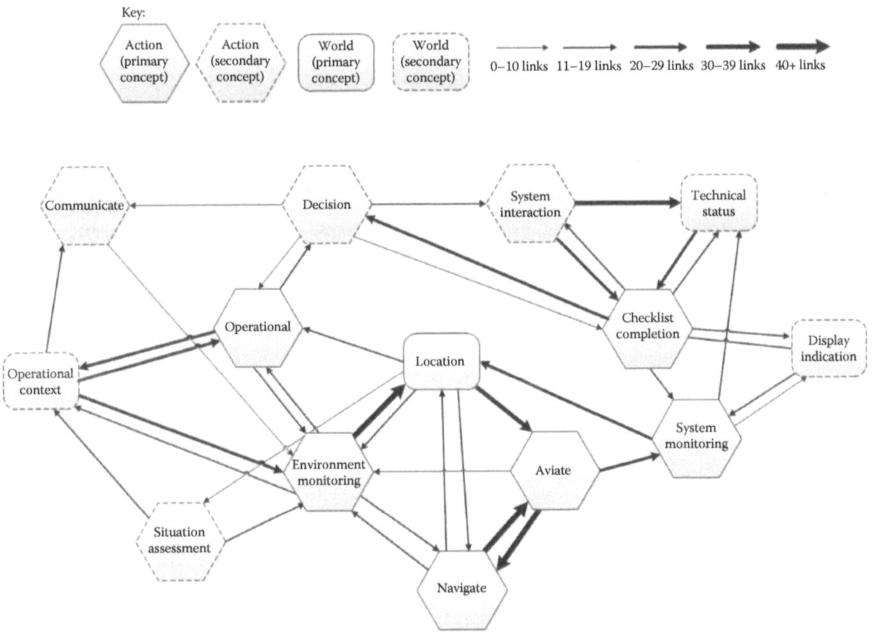

FIGURE 9.5 Generic network model for the SAW-based information network.

concepts were used to create the network model presented in Figure 9.5. The primary and secondary concepts consisted of action and world subtypes, schema subtypes are not represented in Figure 9.5. This aligns with data in previous chapters that found a strong tendency for automatic skill-based behaviour (SBB) characteristic of experts (identified by direct links between action and world concepts and bypassing schema). This will be discussed further in Section 9.5.1.3.

Table 9.12 provides a comparison with the SAW concepts from Chapter 6. Of the 14 primary and secondary concepts, 8 overlap between the two analyses, although the order of importance differs. If the two concepts that were added to this analysis are removed (checklist and operational action), as these could not have been the previous data set, then there is a 66% overlap between the concepts. Aviation and location are primary concepts in both data sets.

The SAW analysis of the SAR data did not take role into account; therefore, two of the case study exercises was explored in more detail using a role-specific SAW taxonomy analysis. Exploring a subset of the data is an approach that has been previously employed by Stanton (2014) in which a simplified information network was created for practical convenience of presenting the metrics, this reasoning holds for the decision to present a subset of the data here. Furthermore, the two case studies are related to exercises with the most tangible decision-making processes and therefore hold most relevance for the overall purpose of this research. Each case study represents approximately 10% of the total data set. For each exercise, an association matrix was built from the relevant from/to links between role/SAW concepts.

TABLE 9.12
Comparison of Importance of SAW Concepts between Individual and Team Data

	Rank Order of Importance (Based on Sociometric Status Value)	
SAW Concept	**Individual (Chapter 6)**	**Team (Chapter 9)**
A_Aviate	1	3
A_Decision	2	12
W_Location	3	4
W_Natural environment conds.	4	18
W_Display indications	5	14
W_Operational context	6	8
W_Physical cues	7	21
S_Direct past exp.	8	23
S_Trained past exp.	9	24
S_Declarative schema	10	22
A_System monitoring	11	5
A_Communicate	12	10
A_Situation assessment	13	13
A_Standard operating procedure	14	17
W_Absent information	15	26
W_Technological conditions	16	11
S_Insufficient schema	17	28
W_Aircraft status	18	16
A_System interaction	19	9
A_Concurrent diagnostics	20	–
A_Non-action	21	–
W_Communicated information	22	15
A_Environment monitoring	23	1
W_Artefacts	24	19
W_Problem severity	25	–
S_Vicarious past exp.	26	27
A_Navigate	27	7
S_Analogical schema	28	25
A_Checklist completion	–	2
A_Operational	–	6
W_Crew status	–	20

9.4.3.3.1 Aborted Winch Training Exercise

This was the first winch training exercise in the SAR training sortie. The crew had arranged to carry out a winch exercise to the vessel as training for both the SAR crew and the vessel crew who were understood to be on a skipper training course. The exercise started with the identification of the vessel and the crew complete a dummy run to assess the available power to the aircraft and winching position.

The aircraft was too high on power and so the crew requested that the vessel speed up, but the aircraft was still too high on power. Coupled with the fact that there were more children on board the vessel than expected, a decision was made to abort the training and look for another winch opportunity.

There were 36 relevant nodes for this exercise. The node-relevant metrics are presented in Table 9.13. The density of the network was 0.06, suggesting a very loosely coupled network, cohesion was 0.01, indicating limited reciprocal links, and diameter was 8, that is 8 'hops' from one side of the network to the other. Using the mean value (0.16) as the cut-off for sociometric status, there were 12 concepts that fell above this value (bold in Table 9.13). These have been used to create the network model presented in Figure 9.6. Again, this aligns with previous data as SBB is displayed in which only action and world concepts are represented. The action of environment monitoring was primarily conducted by the WO and PF for locating the vessel. This updated the PF's world information about the natural environmental conditions and contributed to his task of aviating. During the exercise, the WO communicates with the vessel and in turn this provides communicated information back to the crew which informs the PF's world view of the operational context. The PNF is responsible for conducting the checklist items and the PF makes decisions about each item which leads to the PNF interacting with the aircraft systems. The following exert from the PF demonstrates that the decision made to abort the training was based on a number of factors including world information about the power status and environmental conditions and schemata based on direct past experience of working with children and people's reaction to helicopters:

> We are not really over willing to work with children in close quarters because of the danger. It's about safety, with adults concentrating on the helicopter, people often don't get that close to helicopters and it is noisy, blowing water everywhere. Children are unpredictable; they can't swim as well as adults, that is why we were concerned. Plus with the power, it was consensus not to continue. (PF_interview)

9.4.3.3.2 Confined Area Landing (CAL) Exercise

The second exercise was a CAL and the case study starts from the beginning of the descent after the coastguard had been informed, the landing site had been identified and the briefing conducted in which the pilots discuss the entry strategy, usually considering the mnemonic SBATT: suitability, barriers, approach, touchdown and take-off. After the before landing checks are completed, the initial descent commences and during this time the PF decided to abort the landing and go around because the PNF had bought to his attention that there was a high rate of descent and high power. The aircraft was repositioned and the WO directed the PF to the landing site. For analysis purposes, the exercise ends when the aircraft is on the ground and the landing gear is lowered.

There were 38 relevant nodes for this exercise. The node-relevant metrics are presented in Table 9.14. The network density was 0.06 (loose connections), cohesion was 0.02 (few reciprocal connections) and diameter was 9, that is 9 'hops' from one side of the network to the other. Using the mean value (0.22) as the cut-off for sociometric status, there were 13 concepts that fell above this value

TABLE 9.13
Metric Analysis of the Aborted Winch Training Exercise in Order of Importance High to Low

	Emission	Reception	Sociometric Status
WO_A_COMM.	10	10	**0.57**
PF_A_DECISION	9	9	**0.51**
WO_W_COMM. INFO	8	9	**0.49**
WO_A_OPERATIONAL	7	7	**0.4**
PF_W_NAT ENVIRO	7	6	**0.37**
PNF_A_CHECKLIST	5	5	**0.29**
PNF_A_SYST INT.	5	5	**0.29**
PF_A_ENVIRO MON.	5	4	**0.26**
PF_A_AV.	3	3	**0.17**
PF_A_SIT. ASS.	3	3	**0.17**
WO_A_ENVIRO MON	3	3	**0.17**
PF_W_OP CONTEXT	3	3	**0.17**
PF_S_DPE	2	3	0.14
PF_S_TPE	2	2	0.11
PF_S_DS	2	2	0.11
PF_S_AS	2	2	0.11
PF_A_OPERATIONAL	2	2	0.11
WO_A_SIT. ASS.	2	2	0.11
WM_A_SIT. ASS.	2	2	0.11
PF_W_PHYS CUE	2	2	0.11
PF_W_DISPLAY IND.	2	2	0.11
WO_W_OP CONTEXT	2	2	0.11
WO_W_NAT ENVIRO	1	2	0.09
PF_S_IS	1	1	0.06
PF_A_COMM.	1	1	0.06
PF_A_SOP.	1	1	0.06
PNF_A_SYST MON.	1	1	0.06
PNF_A_ENVIRO MON	1	1	0.06
PNF_A_SIT. ASS.	1	1	0.06
WM_A_OPERATIONAL	1	1	0.06
PF_W_AIRCRAFT STAT	1	1	0.06
WO_W_AIRCRAFT STAT	1	1	0.06
PNF_W_ARTEFACTS	1	1	0.06
PNF_W_DISPLAY IND.	1	1	0.06
WM_W_NAT ENVIRO	1	1	0.06
WM_S_DPE	1	0	0.03

Note: Values in bold are above the mean value.

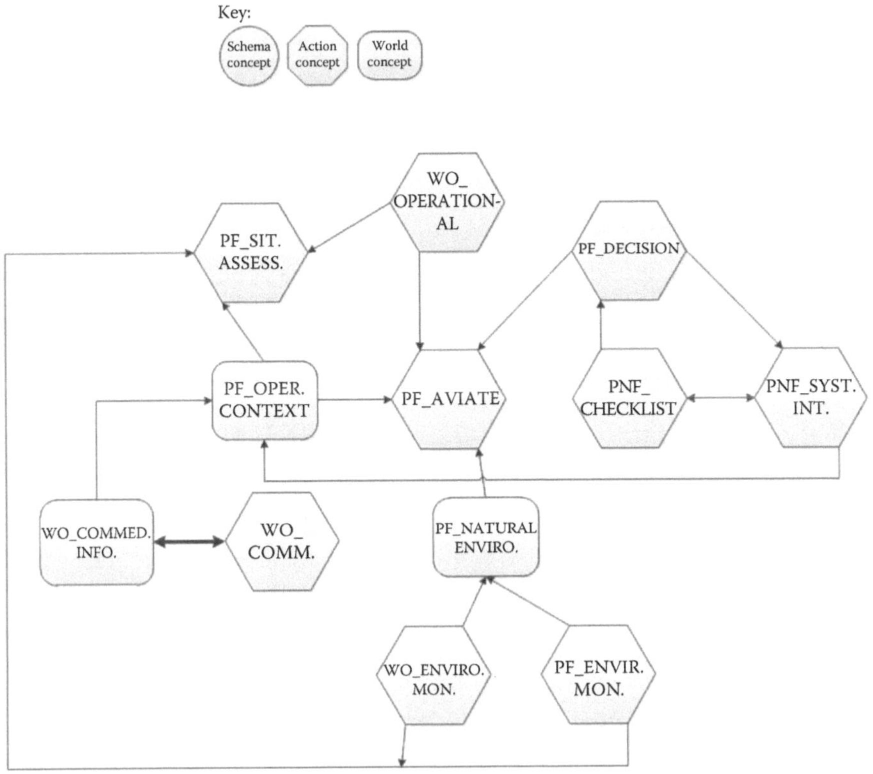

FIGURE 9.6 Network model for SAW concepts and roles for the aborted winch training exercise.

(bold in Table 9.14). These have been used to create the network model presented in Figure 9.7. The CAL exercise is a team activity in which the other crew members act as the eyes for the PF to talk him into the landing site. The landing site was in a clearing in the woods and so the WM and PNF have key roles in monitoring the environment and letting the WO know that the area is clear and the aircraft is a suitable distance away from trees. The WO also monitors the environment and these activities inform the WO's world information about the location, the WO then navigates the PF into the landing site; this navigation action, coupled with the environment monitoring of the other crew members, informs the PF's world information about the landing site and this enables him to manoeuvre the aircraft. This informs the PF's site picture of the landing site; this is a visualisation of the landing area and is akin to an analogical schema. Aside from the navigating/aviating exercise, there is a simultaneous process occurring where the PNF is system monitoring; this is how the high descent rate and high power status was picked up and the action of the PNF provides the PF with world information about aircraft status, coupled with the PF's direct and trained past experiences the situation assessment

TABLE 9.14
Metric Analysis for CAL Exercise in Order of Importance High to Low

	Emission	Reception	Sociometric Status
WO_A_NAVIGATE	33	28	**1.65**
PF_A_AV.	12	17	**0.78**
WO_A_EN. MON.	15	14	**0.78**
WM_A_ENVIRO. MON.	12	13	**0.68**
PF_W_LOCATION	12	13	**0.68**
WO_W_LOCATION	9	7	**0.43**
PNF_A_ENVIRO. MON.	7	7	**0.38**
PNF_A_SYST. MON.	5	6	**0.3**
PF_W_AIRCRAFT STAT.	5	5	**0.27**
PF_A_SIT. ASS.	5	4	**0.24**
PF_S_AS	4	4	**0.22**
PF_A_OP.	4	4	**0.22**
PF_W_PHYS. CUE.	4	4	**0.22**
PNF_A_SIT. ASS.	3	4	0.19
WO_A_OP	3	3	0.16
WO_A_SIT. ASS.	3	3	0.16
WO_W_OP. CONT.	3	2	0.14
PF_S_DPE	2	2	0.11
PF_A_DECISION	2	2	0.11
PF_W_DISPLAY IND.	2	2	0.11
PNF_W_DISPLAY IND.	2	2	0.11
PF_A_SYST. MON.	2	1	0.08
PF_S_TPE	1	1	0.05
PNF_S_VPE	1	1	0.05
WM_S_DPE	1	1	0.05
PF_A_EN. MON.	1	1	0.05
PNF_A_CHECKLIST	1	1	0.05
PNF_A_SYSTEM INT.	1	1	0.05
WO_W_CREW STATUS	1	1	0.05
WO_W_ABSENT INFO.	1	1	0.05
PNF_W_LOCATION	1	1	0.05
PF_W_OP. CONT.	0	1	0.03
WO_W_AIRCRAFT STAT.	0	1	0.03
PF_A_COMM.	0	0	0
PF_A_SOP.	0	0	0
PF_W_NAT. ENVIRO.	0	0	0
WO_W_NAT. ENVIRO.	0	0	0
PNF_W_ARTEFA	0	0	0

Note: Values in bold are above the mean value.

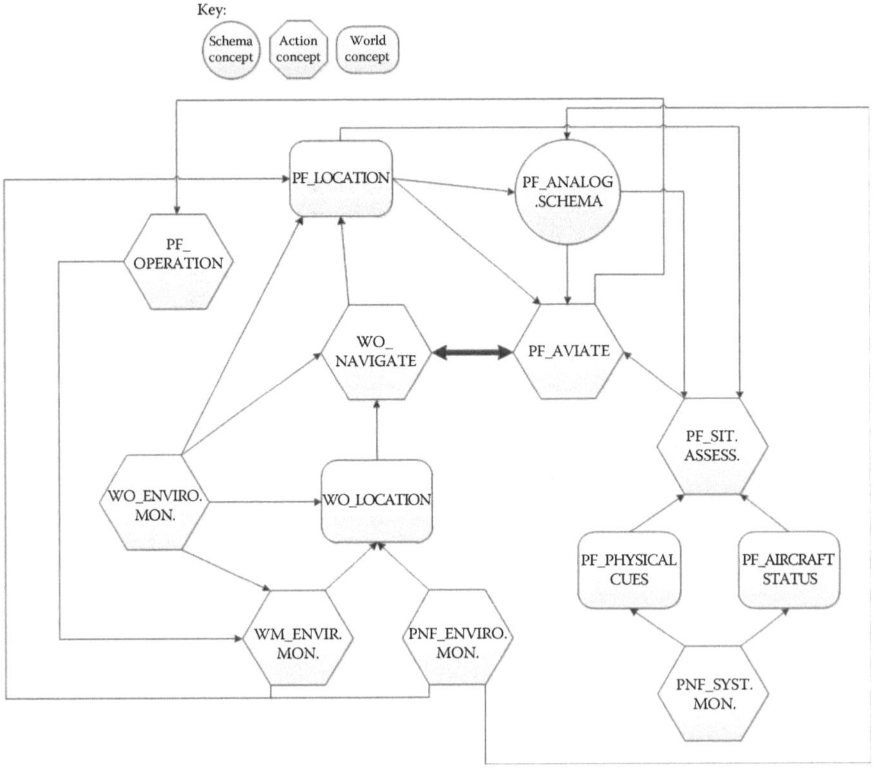

FIGURE 9.7 Network model of SAW concepts and roles for CAL exercise.

led to the fundamental decision to go around. The following text exerts illustrate some of these processes:

> Still clear on the left [WM_ENVIRO. MONITORING]
> And steady, steady, checking what is around [WO_ENVIRO. MONITORING]
> You are clear forward and right PF [PNF_ ENVIRO. MONITORING]
> Roger, moving in [PF_AVIATE] …
> Quite a high rate of descent now [PNF_SYSTEM MONITORING]
> Roger [PF_AIRCRAFT STATUS]
> I knew we would probably have to go around from this based on what the PNF told me about the power and how it felt, this is a standard training routine [PF_TRAINED PAST EXPERIENCE (interview)] …
> You are clear to descend slowly, forward 100 [WO_NAVIGATE].

9.4.4 Communication Analysis

The final analysis looked at the types of communication transactions occurring in the SAR case study. This is not traditionally included in the shortened version of the EAST methodology, but it was felt necessary to gain an understanding of the types

TABLE 9.15
Communication Classification Scheme

Original Category (Bowers et al., 1998)	New Category	Category Description
Uncertainty statements	Query	Direct or indirect task-related questions
Action statements	Instruction	Statement requiring a team member to perform a specific action
Acknowledgements	–	One bit statement following another statement (e.g. 'yes', 'roger' and 'OK')
Responses	–	Statement conveying more than one bit of information (might include an acknowledgement and an instruction)
Plan	–	Statements related to plans and intentions
Factual	–	Objective statement involving verbalisation of realities in the environment
Air traffic control-related	External agent	Statements that were directed to or came from external agents (e.g. ATC, CG, vessels)
–	Judgement	Sharing of information based on subjective interpretation of the situation
–	System message	Communication from systems (e.g. TRAFFIC TRAFFIC)
–	Non-verbal communication	Communication without the use of spoken language (e.g. gestures, body language, facial expressions)
–	Description	Descriptive statements that give an account of what is happening

Source: Adapted from Bowers, C.A. et al., *Human Factors*, 40(4), 672, 1998.

of transactions occurring in the communication data. The data were coded with a communication taxonomy developed by Bowers et al. (1998) specifically for coding aircrew communication data. Some edits were made to the category names and four additional categories were added; the coding scheme is presented in Table 9.15. The categories in the coding scheme align with Kanaki's (1995) assertion that the nature of cockpit communication requires language to issue commands, acknowledge commands, conduct briefings, perform standard callouts, state information, ask questions and convey information. As described in Section 9.2.5, an inter-rater reliability assessment for this classification scheme found very high levels of inter-rater agreement (93%).

Frequency counts of communication types based on role–role transactions were computed, and the percentages of these are presented in Table 9.16. The most represented communication type was acknowledgement (19% of data); within this, the highest number of transactions occurred between the PF to WO (6%), followed by WO to PF (4%). The communication types of factual and description both contained 15% of the data; within the former, transactions between PF and PNF (and vice versa) comprised 5% of the data each, and in the latter, the highest number

TABLE 9.16
Percentages of Communication Types Based on Role–Role Transactions

	Factual	Plan	Response	Query	Instruction	Acknowledgement	Judgement	Description	Ex. Agent	NVC	Syst. Msg	Total
PF-PNF	5	2	2	1	2	5	1	0	0	0	0	18
PF-WO	2	3	2	1	0	6	3	1	0	0	0	18
PF-WM	0	0	0	0	0	0	0	0	0	0	0	0
PNF-PF	5	1	4	2	2	3	1	4	0	0	0	22
PNF-WO	0	0	0	0	0	1	0	0	0	0	0	1
PNF-WM	0	0	0	0	0	0	0	0	0	0	0	0
PNF-EXT. AGENT	0	0	0	0	0	0	0	0	1	0	0	1
WO-PF	2	1	1	4	8	4	1	9	0	0	0	30
WO-PNF	1	0	0	0	0	1	0	0	0	0	0	2
WO-WM	0	0	0	0	0	0	0	0	0	0	0	0
WO-EXT. AGENT	0	0	0	0	0	0	0	0	3	0	0	3
WM-PF	0	0	0	0	0	0	0	1	0	0	0	1
WM-PNF	0	0	0	0	0	0	0	0	0	0	0	0
WM-WO	0	0	0	0	0	0	0	0	0	1	0	1
System-pilots	0	0	0	0	0	0	0	0	0	0	2	2
Total	15	7	10	8	13	19	6	15	4	1	2	100

TABLE 9.17
Extract of Communication Transcript Showing Communication Types and Description of the Exert

Time	From	To	Communication	Type	Description
15:42	WO	PF(C)	*Right, okay, looking at the rotor wash at the moment it is still directly below us*	Factual	The position of the rotorwash means that during a winch training exercise the people on the vessel will get soaked from the rotorwash (spray of water on the sea caused by the helicopter's downwash).
15:44	PF(C)	WO	*Roger…*	Acknowledgement	
15:45	PF(C)	WO	*…Yeah, we are just on 80 at the moment, so we're still not there*	Factual	Eighty refers to the power of the helicopter and 'not being there yet' means that the power is too high for the aircraft to be in a safe single engine configuration to fly away with only one engine if they had an engine failure. Winch training will only commence once safe single engine limits are reached.
15:54	PF(C)	All	*Closing on her at the moment*	Factual	The aircraft is manoeuvring towards the boat and has nearly reached her.
15:55	WO	PF(C)	*Do you want to try 20?*	Query	The WO suggests getting the boat to travel at a speed of 20 knots (it is currently at 15), the increase in speed can reduce the power the aircraft is pulling and therefore might enable safe single engine limits to be reached.
15:57	PF(C)	WM	*Are you happy to go out at that speed WM?*	Query	HEEDFUL INTERRELATING: the PF is checking whether the WM is happy to be winched out at this speed because 20 knots is not the standardised speed to winch from (it is usually 10–15 knots).
16:00	WM	PF(C)	*Yes that's fine*	Acknowledgement	
16:01	PF(C)	WO	*We haven't got the conditions and with the kids can I suggest we go for the ferry? At least see what we've got*	Judgement	The decision is made to abort this winch exercise because there power conditions are not right and also the boat has children on who would get wet from the rotorwash.

of transactions occurred between the WO and PF (9%), followed by PNF to PF (4%). The third most represented communication type, containing 13% of the data, was instruction, and this was dominated by transactions between the WO and PF (8%). Table 9.17 provides a short exert of the communication transcript, specifically demonstrating the process of heedful interrelating which is discussed further in Section 9.5.

9.5 DISCUSSION

Teams are distinguished from groups through the requirement of interdependence; that is, no one person can perform the team's task alone, nor do they have access to exactly the same information as other team members (Jentsch and Bowers, 2005). Stanton et al. (2006, 2009a, 2015) argued that compatible awareness is the process that holds a distributed team together; for effective team performance, there is a need to exchange information among team members. The aim of this study was to explore team perceptual cycle interactions using the shortened EAST method in order to understand how a SAR team function as a distributed cognitive unit. The analysis resulted in information, social and task networks for the SAR training case study; in addition, the type of communication transactions used by the team was also investigated.

9.5.1 Network Interpretations

9.5.1.1 Task Network

The task network demonstrates the linear dependencies that exist in the SAR case study. These exist at two levels. First, they relate to the standardised phases of flight for the whole sortie, where one phase cannot begin without the completion of the previous phase, for example take-off only occurs after taxi. Second, there are linear dependencies within the individual training exercises where the crew must progress through proceduralised checks before they can commence training, for example pre-winch checks, winch brief, dummy run, live run. There is, however, an element of flexibility within this due to the constant assessment of the conditions and their impact on safety. This manifests through the representation of non-sequential tasks for those related to the training exercises. The dependent nature of the overall task network was reflected in the density and cohesion metric values. Contacting the coastguard and transition were the most important concepts in the network (defined by sociometric status), and these are tasks that occur throughout the whole sortie and that other tasks feed into. The other key concepts were primarily those that related to conducting the training exercises, which is unsurprising given the nature of the case study data.

9.5.1.2 Social Networks

The social network analysed the relationships between the agents in the sortie at two levels. First, an all-agent network was constructed. This all-agent network displayed lose coupling within the network, defined by the low density and cohesion scores

(0.2 and 0.15, respectively). Unsurprisingly, the PF(C) was the key agent in this network (defined by sociometric status) and interacted with more individual agents than anyone else. Overall, the five key agents were defined as the captain and co-pilot roles of PF and PNF and the WO. In relation to the pilots, the main role distinction is between PF/PNF (rather than captain/co-pilot) as these are the roles that influence the task allocation and communication transactions with other crew members. Therefore, in subsequent analyses, distinction was not made between the captain and co-pilot. Second, a SAR crew network was constructed (consisting of the four aircraft crew). Compared to the all-agent network, this network shows much tighter coupling between the crew members, demonstrated by the high density and cohesion scores (0.8 and 0.8, respectively), showing a high level of interconnectivity and reciprocal connections between the team members. Rafferty et al. (2012) showed that more effective teams had tighter internal coupling and looser external coupling than less effective teams. The network shows that the primary communication relationships are between PF-PNF and the PF-WO; this was supported by the communication analysis presented in Table 9.16. A comparison of the task and social networks demonstrates how different pictures are provided by each network. For example, in the TNA the task of 'contact coastguard' was the most important concept in the network; however, in the SNA, the coastguard was only ranked 10 out of 15 for importance. This is because although the task occurs frequently not much is actually said and the transaction only occurs with the WO. This highlights an advantage of EAST and its utilisation of complementary networks.

9.5.1.3 Information Networks

The high-level PCM information network shows the interaction between the SAR crew members and the three components of the PCM. The network presented in Figure 9.4 depicts these interactions that represented at least 1% of the data (percentages were low because the text segments were spread over an association matrix with 144 cells, i.e. 12 × 12). Considering the potential size of the network the density value (0.6) suggests some level of connectedness between the concepts; furthermore, the cohesion value (0.5) was indicative of reciprocal links existing in the network which is evident in Figure 9.4. There are clear interactions between different elements of the perceptual cycle and crew members. The network in Figure 9.4 effectively combines the social network with the PCM elements; in doing so, it provides further information about the nature of the interactions between the crew members that the social network alone does not reveal. For example, there are strong links between the actions of the WO and PNF informing the world information of the PF. The communication analysis sheds further light on the nature of these interactions; for example, the two main types of communication from the WO to the PF are description and instruction (see Table 9.6). During the CDM interviews conducted for Chapter 7, the WO stated:

> …because the pilot can't see where we are going to put the man, he can't see the man on the wire, I'm his eyes, so I'm telling him…describing the situation, telling him the distances he's got to fly, whether he's too high, too low and making sure all the obstructions like a cliff, whatever we are working with, is clear. So we've got a clear run, he

> can't see that, he can't see down. So I'm talking him in, where the winchman is, there is a pattern, it just flows... So really I am passing instructions to him but then at the same time my description is what he sees. I'm his mental model. (WO with over 6000 role-specific hours at the time of interview)

This was certainly evident in the communication transcripts from the WO in the current case study. The WO describes factors such as the location of vessels, the surrounding environment and tasks that he is performing that the PF cannot see; the text excerpts exemplify this:

> Okay, the target is just to the right of the aircraft's nose in about our half past 12 position...
>
> Okay, just having a little check back and left, you've got a yacht which is keeping station with us in our 8 o'clock and about 50 units, so don't go too far to the left when we come away
>
> ...Getting [name of WM] outboard, out the door, Winching out...

The WO also has a role to instruct the PF to manoeuvre the aircraft towards targets that the PF is not usually visual with, for example:

> Forward and right three as we approach the side of the vessel...Forward and right two.

In return, the more frequently used communication type from the PF to the WO was 'acknowledgement'. These communication transactions are examples of shared leadership displayed by the SAR team, whereby any team member, not just the formal leader, can take on leadership functions depending on the situational requirements, member competencies and resources (Grote et al., 2010). Overall, the analysis of types of communication is consistent with Bourgeon et al.'s (2013) finding that aircrew communication is characterised by the sharing of information and the directing of actions.

The analysis with the SAW taxonomy demonstrated the specific manifestations of the three elements of the PCM. Focus will only be given to the primary and secondary concepts as these are most relevant to understanding the team perceptual cycle interactions. These concepts consist of action and world subtypes, rather than schema. The dominant role of action and world concepts has been consistently found in other chapters (see Chapters 5 and 6) and has been attributed to the SBB characteristic of experts. In SBB, sensory information in the world is processed at the skill-based level in which information is acted upon automatically (Rasmussen, 1983). In the SAR team, the sensory input may come from different team members but they do not have to be interpreted or integrated but just responded to. The SAW model presented in Figure 9.6 was developed with data from expert crews and therefore behaviour is largely automatic, hence the bypass of the schema concepts. As in previous chapters, the data collection method may have also influenced the findings, and this is discussed in Section 9.5.3.

Table 9.11 compares the primary and secondary concepts between the concurrent team data analysed in this chapter and the retrospective individual data in Chapter 6. There is a relatively high degree of overlap (66%) of key concepts represented in the two data sets, and the action concept of aviate and the world concept of location are primary concepts in both data sets (although the order of importance varies). Differences between the concepts and their order of importance reflect the nature of the data sets; for example, the key concepts in this chapter reflect the nature of the SAR training exercises in a team context (e.g. environment monitoring, checklist completion, operational action, navigate), whereas the key concepts in Chapter 6 are more relevant to dealing with critical incidents (decision, display indication, physical cues). These findings provide a provisional concurrent validity assessment of the SAW taxonomy for gaining a detailed understanding of perceptual cycle interactions in aviation. A key difference between the two data sets is the role of schema; in Chapter 6, three different types of schemata were identified as secondary concepts (direct past experience, trained past experience and declarative schema); however, in the SAR analysis, there were no schema subtypes in the primary or secondary concepts and they only feature towards the lower end of the tertiary concepts. This is likely to reflect the differences between the two data collection methods that were employed and is discussed in more detail in Section 9.5.4.

The two SAW analyses of the specific training exercises reveal the role specific interactions of the SAW taxonomy concepts. The results demonstrate the interaction of the different elements of the SAW taxonomy and how the actions of one crew member (e.g. PNF_system monitoring) provide world information for other crew members (e.g. PF_aircraft status) and how this influences further interaction in the environment (e.g. PF_situation assessment). Wilson et al. (2007) highlighted the importance of mutual performance monitoring and back-up behaviour to avoid team work breakdown, and this was evident in the CAL exercise when the PNF alerted the PF to the high rate of descent and power and ultimately a decision was made to go around.

9.5.2 Gaining Insights into Distributed Cognition in the SAR Crew

Stanton (2014) argued that the aim of methods to analyse distributed cognition is to make the complexity of socio-technical systems more explicit so that interactions can be examined and to reduce this complexity to a manageable level. The EAST method has been proposed as a suitable method to achieve these aims (Walker et al., 2010; Stanton, 2014). Furthermore, it is acknowledged that it is important to study team processes rather than outcome; however, these are harder to quantify and present more of a research challenge (Annett and Stanton, 2000; Paris et al., 2000). Applying EAST in conjunction with a PCM analysis is able to demonstrate how teams function as an information-processing unit, that is how knowledge is acquired, distributed and acted on (Paris et al., 2000b).

The analysis presented here enables insights into the distributed cognition of the SAR team to be understood: The task network showed what was done; the social networks showed which team members were most connected, the information networks demonstrated how these connections manifested as perceptual cycle interactions and the communication analysis showed what types of transactions connected the team members together. One of the key findings of this analysis was the difference between the all-agent social network and the SAR crew social network.

The all-agent network displayed loose coupling between the agents, whereas the SAR crew network showed much tighter coupling demonstrated by a highly dense and cohesive network. This supports Rafferty et al.'s (2012) assertion that in small, co-located teams, there needs to be a tightly integrated set of information elements (i.e. tight coupling in the SAR crew network), whereas in larger distributed teams this is less important because everyone does not need to know everything (i.e. loose coupling in the all-agent network). In terms of DSA, it would be inappropriate and a waste of resources for everyone in a distributed team to know everything because of the diversity of roles within the all-agent team; for example, ATC do not need to know when the SAR crew speak to the CG. A tightly coupled team, however, is indicative of strong connections and reciprocal links between team members over which information can be shared. The SAR context warrants a particularly tight coupling between team members, especially in relation to the PF and WO, whereby the WO acts as the eyes for the PF, providing him with essential information that the PF cannot access himself. Similarly, in an analysis of decision-making in a joint fires (Army and RAF) mission training exercise, Rafferty et al. (2012) found that higher levels of density and cohesion were apparent in the more, rather than less, effective team.

Stanton et al. (2010) warned that one of the disadvantages of distributed teams is that tightly coupled work, where individual tasks are highly dependent on the work of others, is more difficult. One potential solution to this is cross-training whereby each team member is trained for the role of other team members (McCann et al., 2000). This interpositional rotation was actually evident in the SAR crew whereby the PF and PNF switched roles within the training sortie. It is also often the case that within SAR crews the WO used to be a WM but changed role at a certain age or physical condition. McCann et al. (2000) found that cross-training did not benefit outcome measures of team performance such as speed and accuracy, but it did positively impact process measures of team performance. This book has consistently argued that the application of the PCM, and the associated consideration of DSA, is concerned with understanding processes rather than outcome. Clearly, there are operational limitations associated with the practicalities of cross-training, but it is a potential recommendation from research of this nature.

The appreciation of other roles evident in the SAR crew is likely to support the high-functioning team processes demonstrated by them. For example, the SAR team displayed evidence of the three characteristics of advanced and effective team interaction as defined by Grote et al. (2010). In terms of implicit coordination, which occurs when the team have compatible awareness of task requirements, a frequently used indicator of this is the provision of unsolicited information, that is delivering task relevant and timely information without being asked to do so, this is evident throughout the case study and is exemplified next:

> I think we'll go around from that one. (PF)
>
> Okay, you are well clear below, clear to transit away. (WO)

The verbalised decision of the PF to go around from the initial CAL was immediately followed by the WO determining that the area was clear without being asked. This can be linked to the level of standardisation of the team environment; rules and procedures are an important information source for developing a compatible understanding of the situation. The SAR team environment is highly standardised, demonstrated by the sequential dependencies exhibited in the task networks. High levels of standardisation are linked to more implicit coordination because the reliance on a shared set of standards directs the actions of the team in following the known conventions. This also has an impact on the second characteristic; heedful interrelating, that is deliberate efforts made by team members to consider the effects of their actions in relation to others. Grote et al. (2010) argued that this trait characterises a team functioning as a distributed cognitive unit because it depends on a sufficiently compatible understanding of the situation, but conversely, heedful interrelating can also result in increased levels of compatible awareness by pooling information, knowledge and experience and therefore is a mediator of compatible SA. There is, however, less heedful interrelating required in a highly standardised environment because the reliance on known conventions reduces the need to explicitly explore other team members' perspectives (Grote et al., 2010). This was evident in the SAR sortie whereby implicit coordination dominated the communication transactions unless the situation was less standardised than expected and then heedful interrelating was evident and is exemplified in the text exerts provided in Table 9.17. The final characteristic is shared leadership, and as previously described, this is displayed in the SAR team, evident by the fact the WO is the second most important agent in the social network and in the authoritative communication types displayed (e.g. instructions).

9.5.3 Evaluation of Crew Teamwork

Grote et al. (2010) criticised previous research on team processes for being too inductive and conducted without reference to a broader theoretical framework; arguing that that research founded in theory often provides a more systematic approach and therefore results in more useful practical recommendations. Furthermore, it has been argued that the main weakness in the analysis and understanding of distributed cognition is the complexity of its measurement and representation (Salmon et al., 2008; Stanton et al., 2009b). The research presented here took a deductive approach by conceptualising team interactions in the PCM framework and applied the shortened version of EAST to this analysis; this is a relatively simple way for analysing the SAR team from the perspective of distributed cognition. However, a disadvantage of the shortened version of EAST is apparent when the role of the WM is considered. The WM is not a key agent in the social network, but from the description of the SAR roles in Section 9.2.1, it is an integral role in the SAR team, being the one who is winched out and coordinates the rescue. The networks in the shortened version were built from communication data only, whereas the traditional version relies of constitute methods of HTA, CDM, coordination demand analysis, communication usage diagrams and operation sequence diagrams (Stanton, 2014). The advantages of the shortened version in terms of saved time and resources outweigh any limitations, provided the analysis is interpreted with the acknowledgement of potential limitations.

The comparison of the SAW taxonomy analysis with the analysis in Chapter 6 has demonstrated that the taxonomy can be applied to data collected from communication transcripts, and the amendments to it means that it is now appropriate for the analysis of team perceptual cycle interactions and is therefore a more comprehensive taxonomy. In Chapter 6, it was recommended that the SAW taxonomy needs to demonstrate usefulness as an aid in interpreting and integrating research results as well as establishing whether new data sets can be accurately assigned to the categories. The findings in this chapter have begun this process, and from this initial analysis, it would appear that the SAW taxonomy provides a valid way of interpreting different data sets; this is particularly reflected in the differences between the key concept types and the order of importance in the overlapping concepts as it demonstrates that the SAW taxonomy is sensitive to data collected from different sources (i.e. retrospective interviews vs. concurrent verbal communication) and in different contexts (i.e. critical decision-making vs. training sorties). Furthermore, the findings have supported the assertion presented in Chapter 6 that the taxonomy could be applied to non-critical decision-making data. Again, this is reflected in the relative importance of concepts because the three concepts that did not appear in this data were the ones more relevant to dealing with critical incidents (i.e. concurrent diagnostics, non-action and problem severity).

Chapter 8 introduced the notion of network representations which has been further developed in this chapter and, in doing so, has enhanced the EAST method in three key ways. First, the application of the PCM coding scheme and the SAW taxonomy to generate the information networks offers a more generalisable level of analysis. Traditionally, information networks are generated from key words in the raw data; however, by coding these text segments to a higher level of abstraction means that the networks are more generalisable and therefore have a higher utility. Network analysis has often been used to make comparisons between groups or conditions. For example, Walker et al. (2011) used networks to explore the cognitive compatibility between motorcyclists and car drivers. Similarly, Salmon et al. (2014a,b) used networks to explore differences in SA between drivers, cyclists and motorcyclists. These studies used automatic network generation software tools whereby network differences are represented by the presence or absence of concepts across the different road user networks. If, however, the data were coded with a standard classification scheme, such as the SAW taxonomy, comparisons could be conducted with reference to the PCM. Whilst this approach requires an additional phase of thematic analysis, the advantage is that a more generic network could be produced which could aid comparisons between diverse data sets.

The second enhancement to EAST is through the combination of social and information networks by the creation of role-specific association matrices. Previously, studies that have explored the interaction of different networks have conducted an additional phase of analysis to combine them. For example, Stanton (2014) used a colour coding system to analyse the task and information networks from the perspective of the social networks. The nodes in the networks presented in Figures 9.4, 9.6 and 9.7 contained both role and concept information and therefore automatically account for both the social and information factors. This approach is supported by Sexton and Helmreich's (2000) observation that the dynamic and interdependent nature of cockpit processes means that collapsing data across positions dilutes the effects of individual roles.

Finally, the addition of the communication classification scheme added a new layer of analysis and potentially a fourth phase in the shortened EAST method, or a precursor to the construction of the task, social and information networks. As described in the introduction, communication is the foundation of team performance (Jentsch and Bowers, 2005); therefore, it does not seem appropriate to explore team interactions without some consideration of communication processes, particularly when data are collected from communication recordings. The communication analysis provided further insights about the nature of the perceptual cycle interactions in the SAR team, although this analysis was rather rudimentary in that it only computed frequency counts of types of communication transactions. There is general consensus in the literature that frequency counts alone are not adequate to gain a complete understanding of team communication processes, rather patterns in the data should be explored (Kanaki et al., 1991; Sexton and Helmreich, 2000; Jentsch and Bowers, 2005; Grote et al., 2010; Bourgeon et al., 2013). The EAST methodology could be enhanced with the formal inclusion of a communication analysis network phase; an association matrix of communication transactions could be created and subjected to the same analysis of metrics as the other phases.

The final point for consideration in this section concerns the data collection process and its impact on the results. The schema category is under-represented in this data set (none of the primary or secondary concepts related to schema subtypes), and although this has been discussed as an issue throughout the book, the data collection method of communication recording appeared to yield proportionally less schema-based data than the CDM; the data set in Chapter 8 (team CDM interviews) contained 27% schema-based data; however, the data set in this chapter contains only 3% schema-based data. This research sought to explore the three individual elements of the PCM, and therefore, these were used as concepts to construct the networks; as such, the networks represent the whole perceptual cycle process (including the schema, action and world elements within them). Other researchers have taken a different approach and used the entire network as a representation of the phenotype schema (see Stanton et al., 2009a,b; Walker et al., 2011; Salmon et al., 2014). However, this approach does not account for the interacting elements of the PCM and would therefore not have achieved the aims of this research; although it could be an interesting avenue to explore especially in light of limited schema-based data.

There were advantages to the concurrent communication recording method insomuch that the criticisms levelled at the CDM about self-report and memory degradation are overcome. A different data collection method also enables the validity of previous findings (e.g. SAW taxonomy) to be considered. As highlighted in

Chapter 2, schemata are not observable and therefore can only be inferred through their behavioural manifestations. Schematic processing would have underpinned each crew members information processing; the inherent nature of schemata is highlighted later in which Greiser et al. (2014, p. 174) described the pilot's perspective of the helicopter take-off procedure:

> ...the pilot catches up on the mission and checks the weather and notice to airmen. If the take-off is possible in accordance with the flight manual and operational restrictions, the pilot takes a seat in the helicopter and goes through a mental checklist. Beside technical matters, they analyse the initial heading with respect to the wind direction and wind speed, position of the sun...the confinement of the area, the terrain, the emergency landing spots, and areas that should be avoided due to noise restrictions. Then the pilot decides on a take-off direction and procedure and starts the engines...

In accordance with Neisser's (1976) PCM, it is the activation and utilisation of schemata that enables the situation assessment to occur; the pilot holds schemata about weather conditions, operational limits and the environmental assessment. As described in the SAW taxonomy, these schemata are a result of factors such as direct past experience, training and declarative knowledge. However, because the use of schemata is so implicit, it is rarely verbalised in communications between team members, hence their under-representation in this data set. As such, a multi-method approach is recommended, for example concurrent observations/communication recording with debrief interviews. This affords the benefit of the concurrent data collection and should enable the verbalisation of schema-based data, particularly if the schema section of the SWARM interview schedule is used (see Chapter 6). Alternatively, concurrent verbal 'think-aloud' methods may yield more schema-based data, particularly in relation to phenotype schema which are defined as the activated responses in a given situation and can therefore be inferred through data collection methods (Stanton et al., 2009b). This method, however, is associated with task performance issues, and it would not be possible to conduct in a naturalistic aviation study. The results do highlight that the thematic analysis was as unbiased as possible and entirely data driven, that is assumptions were not made about schemata. The naturalistic nature and ecological validity of the data collection cannot be disputed and as with any research approach trade-offs will be evident.

9.5.4 Implications and Avenues for Future Research

In recent years, the DSA approach has made progress in understanding the manifestation of SA in teams (Stanton et al., 2015). There is consensus in the literature that communication and coordination are the most important constructs to study in order to understand successful team performance within a distributed cognition framework (Artman and Garbis, 1998; Grote et al., 2010; Rafferty et al., 2010). Combining the PCM (including the SAW taxonomy to provide a detailed level of abstraction) with the EAST methodology (including a communication taxonomy) has the potential to provide a unifying and common framework to the analysis of team processes.

A number of avenues for future research have been highlighted by this study with associated implications and the potential for practical application. As alluded to in the previous section, the EAST methodology could be enhanced with the inclusion of a communication network analysis phase, particularly as research has consistently found that communication is a key aspect of DSA (Rafferty et al., 2012, 2013; Sorensen and Stanton, 2013). Furthermore, Jentsch and Bowers (2005) argued that frequency counts alone are not enough to understand communication processes. The usual purpose of communication analysis is to make comparisons between effective and poor teams to establish how communication differences manifest (Bowers et al., 1998; Sexton and Helmreich, 2000; Jentsch and Bowers, 2005; Grote et al., 2010; Bourgeon et al., 2013). Applying the principles of EAST and the comparison this affords could be a novel and insightful way to analyse team communication and feed this into deductively established training requirements (Grote et al., 2010).

TDM training is designed to amalgamate individual skills into a team capability; however, Salas et al. (2007) argued that few TDM programmes are based on solid theoretical underpinnings. As argued throughout this book, the PCM provides a strong theoretical perspective from which to understand decision-making. This chapter has demonstrated that the PCM and DSA together can account for team interactions and explain the processes involved in reaching compatible awareness, and therefore, there is potential utility in structuring training around these processes, such as the cross-training approach previously discussed. These ideas are further described in detail in Chapter 10.

Research of this nature also has relevance for the design of adaptive displays. Stanton et al. (2009) argued that when designing with a distributed cognitive system in mind, then displays that are designed to support team functions must be able to cater for the different roles of team members. Understanding the interaction of different information elements between different team members can aid designers in creating functional displays that support how different pieces of information are linked and utilised by different team members which would translate into how these are grouped and presented on an interface. This is discussed further in Chapter 10.

9.6 CONCLUSIONS

This chapter sought to explore team perceptual cycle interactions in data collected from concurrent communication recordings. The PCM provides a distributed cognition perspective and allows the acquisition and maintenance of compatible awareness within a team to be considered. A further aim of the research was to investigate the validity of the SAW taxonomy that was developed in Chapter 6. To achieve these aims, the EAST method was employed because it acknowledges the complexity within team interactions and provides multiple perspectives to appreciate the relationships between different team members, their tasks and their information requirements (Stanton, 2014). This approach has allowed insights to be gained about team perceptual cycle interactions, showing how the SAR team functions within a distributed perceptual cycle, whereby the actions of one team member become world information for another team member. In doing so, the EAST method has been advanced in three key ways:

1. Utilising the SAW taxonomy in the information network phase creates more generalisable networks that could be used for comparison purposes.
2. Combining the social and information network analysis by role-specific node coding removes the need for an additional analysis phase and is more sensitive to role-relevant information than an colour coding approach might be.
3. Adding the communication coding sheds more insight on the types of team interactions occurring and could be further advanced to be the fourth phase of EAST.

Sexton and Helmreich (2000) argued that compatible awareness is imperative for effective teamwork and communication to be realised in the cockpit. This research has demonstrated how communication and the way in which a team function within a perceptual cycle enable this compatible awareness to be achieved. The shortened EAST method provides a relatively efficient way to analyse and graphically represent distributed cognition (Stanton, 2014). The analysis presented here was concerned with the immediate aircraft crew; however, there are no boundaries to the analysis, meaning it is possible to apply it to complex socio-technical 'systems of systems' using a 'networks of networks' approach.

This study has continued to advance the consideration of team perceptual cycle interactions that were introduced in Chapter 8. The network approach is capable of modelling the interconnectivity that exists between team members, and with the advances suggested to the EAST method, it is hoped that research will continue in this vein to further our understanding of team processes. The importance of this from a SAR perspective is summarised by a WM, but the worlds resonate for any team operating in a safety-critical system:

> Safety is, of course, our top concern. We always have an in-depth brief and debrief, we discuss the exact things we intend to do. Everyone has to know what everyone else is doing at ALL times. You have to be so in tune with each other.
>
> **Bond Aviation Group (2015)**

The book is concluded in Chapter 10, which summarises the research objectives in light of the findings and discusses avenues for further enquiry that have arisen as a result of this research.

10 Conclusions

10.1 RESEARCH SUMMARY

The aim of the research presented in this book was to investigate the application of Neisser's (1976) perceptual cycle model (PCM) to study the process of aeronautical decision-making (ADM) both in individuals and teams. This aim encompassed validating and advancing the explanatory power of the PCM. The main findings are summarised later, including a discussion of the contributions and associated implications of the research. The work presented in the book was structured around three key objectives, and the findings are summarised in relation to these objectives.

10.1.1 Objective 1: Understand the Role of the Perception–Decision–Action Cycle in ACDM

This objective was initially achieved through two literature reviews: the first (Chapter 2) detailed the role of Schema Theory in Ergonomics research. Schema Theory is a central tenant of the PCM as Neisser (1976) proposed that it is these internal mental templates organise our world knowledge and guide interpretation of externally available information. The active nature of schemata means that they can be modified and updated by environmental conditions; this reciprocal, cyclical relationship between schemata and the environment forms the basis of the PCM. A review of the literature highlighted that Schema Theory has proved contentious since its emergence in the literature. This is primarily because of the mentalist nature of schemata, and, therefore, Schema Theory is constrained by our ability to determine what a schema actually is. Schema Theory has been accused of not being a true theory. However, in Chapter 2, it was demonstrated that Schema Theory does fit the generic requirements of a theory as defined by Kerlinger and Lee (2000), namely that a tentative explanation of reality is offered that generates testable hypotheses and is empirically validated to establish how accurately the theory does represent reality. It was also found that Schema Theory has been successfully applied in a number of key Ergonomics research areas, including the study of naturalistic decision-making (NDM), human error and situation awareness.

The second review (Chapter 3) applied the principles of Schema Theory and the PCM to understanding the pilot actions that led to the Kegworth plane crash. A common theme to emerge in the literature was that labels such as 'human error' or 'bad decision-making' do not provide an adequate explanation of incidents or accidents. Traditionally accident or incident investigations concluded with these labels, whereas a 'new view' prevails in which these labels are the starting points for uncovering causal processes. It was found that the decision-making actions of the pilots could be explained by considering the situational information they were exposed to and the schemata they held for the various situations they found themselves in.

The research demonstrated that utilising the PCM as the theoretical framework for accident analysis addressed the five requirements necessary to achieve this understanding of decision-making.

A second case study was provided in Chapter 4; this analysed data collected about a critical incident via the critical decision method (CDM). The analysis demonstrated that the ACDM process was underpinned by the elements depicted in the PCM, namely internal schemata and external environmental information, offering further support for the use of the PCM as the theoretical perspective to explore ADM.

The PCM has been utilised as the theoretical underpinning for the research presented in this book. There has not been any evidence of the validity of the model presented in the literature. The PCM assumes that information processing occurs in a unidirectional cycle, but this has not been verified. Therefore, Chapter 5 sought to determine whether the unidirectional flow of information was an accurate representation. Data from CDM interviews were qualitatively analysed using the high-level PCM coding scheme. It was found that the traditional cycle accounted for the majority of the data, but also a counter-cycle existed, particularly in a reciprocal mini-cycle between the action and world components of the PCM. This finding was also supported with data collected via team CDM interviews and team communication transcripts. This counter-cycle was explained from the context of Rasmussen's (1983) Skills Rules Knowledge taxonomy. It was argued that the reciprocal cycle between world–action–world was akin to the automatic, skill-based, behaviour characteristic of experts. Overall, it was determined that the PCM provides an accurate representation of ACDM; however, consideration of the counter-cycle can add to its explanatory power. This is a novel finding and presents an important extension to the PCM. Demonstrating the direct link between the world and action components of the PCM suggests that different levels of information processing occur within expert behaviour, which may be dependent on the person or the situation. Additional research is recommended to further validate this finding. The researchers and practitioners that use the PCM in their research need to consider the counter-cycle in their analyses and explanations because the perceptual cycle process may not be as clear-cut as previously thought.

Subsequent research explored the three elements of the PCM in more detail (Chapter 6). Via a combination of thematic analysis and social network analysis (SNA), it was found that ADM consists of several core concepts and the importance of these vary depending on the phase of critical incident decision-making. Overall, the core action concepts including 'aviate' and 'decision action' and 'situation assessment' were associated with schema concepts such as 'direct past experience' and 'trained past experience' and world concepts including 'display indications' and 'natural environmental conditions'. The results were used to develop a network model of ADM, presented in Chapter 6.

10.1.2 Objective 2: The Development of Elicitation and Analysis Methodologies

This objective was addressed by the iterative development of a coding scheme to analysis qualitative data using a thematic analysis processes. The first level of the coding scheme enabled deductive thematic analysis and was based around Boyatzis' (1998) criteria for developing a meaningful coding scheme (Chapter 4). The three key elements of the PCM, namely schema, action and world, were used as the basis for coding critical incident data. The findings highlighted that, whilst perceptual cycle processing could be understood, the explanation was at a very high level (i.e. schema, action and world), and this limited the descriptive power of the model. As a result of this, the PCM was considered at a more detailed level of abstraction and the schema–action–world (SAW) taxonomy was developed through inductive thematic analysis, in which categories (or subtypes) of three PCM elements were defined and used to determine the key concepts involved in ADM (Chapter 6). A novel approach to pair the qualitative data analysis with the computation of SNA metrics (particularly sociometric status) was employed.

The reliability of the coding scheme has been consistently established. A review of the literature determined that percentage agreement was the most appropriate method for assessing the reliability of the coding scheme and various studies of inter- and intra-rater reliability were conducted. These studies consistently found agreement of 80% or above and, therefore, there is confidence that a reliable coding scheme has been created. Additionally, the retest reliability of the CDM was assessed (Chapter 4). It was found that, even after a time lapse of over two years, the method allowed for the reliable elicitation of critical decision-making data.

The application of the CDM also identified some flaws with the method as a way of eliciting perceptual cycle data. As such, the SAW taxonomy was used to develop cognitive prompts in an interview schedule called schema–world–action research method (Chapter 7). The validity and reliability of this approach was determined via a critical incident interview. It is anticipated that SWARM will provide a domain-independent method for eliciting and understanding perceptual cycle information.

In Chapter 9, the SAW taxonomy was refined to make it more appropriate for the analysis of team data; it was edited to reflect categories relevant to team data and the results of the analysis provide an initial assessment of the validity of the taxonomy. Furthermore, for the team data, the qualitative data analysis was combined with the EAST methodology to provide a more robust analysis of perceptual cycle processes; in doing so, enhancements to the EAST methodology were proposed. The application of a communication coding scheme also enhanced the insights gained from the SAW taxonomy analysis of team data.

10.1.3 Objective 3: Investigate How Teams Interact in the Perception–Decision–Action Cycle

The PCM is an individual model of cognition; however, distributed decision-making by teams of people characterises the NDM environment in most socio-technical systems and aviation is no exception. The PCM has been used as the theoretical underpinning to explain team processes such as distributed situation awareness (Stanton et al., 2006; Salmon et al., 2008), although team perceptual cycle processes have not been explicitly investigated. The final research objective was to explore team perceptual cycle interactions. This was grounded in the SAR context because the search and rescue (SAR) aircraft team present a self-contained unit of analysis and operate in a classic NDM environment.

The CDM was utilised in a team context and the data analysed using high-level PCM coding scheme (Chapter 8). It was found that the traditional depiction of the PCM was not suitable to model the team interactions that occur and a network representation was found to be more appropriate. A second study utilised a different data collection approach; concurrent communication transcripts (Chapter 9). This data set was used to assess the external validity of the SAW taxonomy, and, as a result, edits were made to the classification scheme. It was found that the SAR team function in a distributed perceptual cycle, with the actions of one team member becoming world information for another team member. The EAST method was employed to analyse and represent team perceptual cycle interactions, and, as a result of the research, three enhancements to the method were proposed.

10.2 FUTURE RESEARCH

The potential implications of the research presented in this book have raised several lines for further academic enquiry. They are structured around three key areas of theoretical, methodological and practical implications of this research.

10.2.1 Theoretical Implications

Kurt Lewin stated *there is nothing so practical as good theory* (cf. O'Hare et al., 1998); training and design solutions require the considered integration of theory and practice. This book presents efforts to advance the perceptual cycle model of information processing by assessing its validity and deconstructing the constituent parts into a more detailed level of abstraction. However, further research efforts can continue to advance the theoretical underpinnings of the PCM.

10.2.1.1 Diverse Domains and Demographic Groups

The work presented in this book is based around the aviation domain; however, the research focus of exploring individual and team decision-making from the theoretical perspective of the PCM is domain independent and has potential applicability in a variety of domains involving critical decision-making tasks. As discussed in previous chapters, the PCM has been applied in other transport domains including rail and road research (e.g. Stanton and Walker, 2011; Salmon et al., 2013a, 2014a,b). It would be interesting to explore the application of some of the developments and findings in this research to these and other domains. The current specificity of the SAW taxonomy has been recognised in Chapter 6, and it has been suggested that the domain-specific categories (e.g. aviation and aircraft status) could easily be supplemented by the domain under analysis as the descriptions for these categories are general enough to apply to any system interaction, particularly other transport domains. This research, however, is certainly not limited to only transport domains, and it is likely to have relevance in any safety critical domain where the potential for critical decision-making exists. For example, in a review of the use of classification schemes in healthcare Mitchell et al. (2014) advocated the need for classification schemes to take account of the combination of cognitive and environmental factors (i.e. as considered in the SAW taxonomy). Other domains of potential application include the oil and gas sector and the nuclear power industry.

A further area for investigation is exploring the findings of this research in relation to different demographic groups. For example, Greiser et al. (2014) found that whether helicopter pilots were from a civilian or military background influenced their flight planning process. An interesting comparison in this research would be between novices and experts. Schemata are developed through past experience and expectations, and therefore expert decision makers will have a larger repertoire of schemata to draw upon to enable them to select relevant environmental cues and therefore be more selective than novices in the information they attend to and this enables the skill-based behaviour that is characteristic of experts (Paris et al., 2000b). No distinction was made in this research between expert and novice pilots, and it would be interesting to determine how the expected perceptual cycle processing differences manifest between these two demographic groups. This would also contribute to the verification of the SAW taxonomy; Fleishman and Quaintance (1984) argued that a comprehensive taxonomy should be capable of distinguishing between different populations, and therefore the SAW taxonomy should be capable of highlighting differences between experts and novices.

10.2.1.2 Predictive Utility

The current research has focused primarily on the retrospective exploration of individual ACDM and the concurrent exploration of team perceptual cycle processes. It would be interesting to explore whether there is predictive power in the approach, for example in predicting where potential weaknesses in ACDM might lie at both the individual and team levels. There has been a call in the aviation literature for proaction rather than reaction with regard to safety (McFadden and Towell, 1999; Maurino, 2000). Similarly, Hancock et al. (2009) argued that predictive knowledge is the key to making the next step in scientific progress; without being able to predict behaviour, behavioural modification will not be possible. Chapter 2 demonstrated that Schema Theory, and therefore by association, the PCM, allow for predictions to be made. Furthermore, Griffin et al. (2010) explored the predictive value of network modelling in aviation accident analysis. It was found that 'ideal' and 'actual' scenarios could be compared to show differences in communication links and the use of information between them and in turn this allowed for the identification of routes into the system for misinformation, false alarms or missed information. Future research could advance this approach specifically for ACDM, combining the PCM, Schema Theory and network modelling to help predict what *can* go wrong based on a description of what *should* happen. This will enable the identification of potential areas of vulnerability that exist for decision makers, without an incident or accident occurring, thus advancing a proactive approach to the study of ADM. The use of schema-based theories for developing predictive models could be criticised for being too specific; the very nature of schemata means that they are unique and personal manifestations of knowledge. However, Hancock et al. (2009) argued that prospective theories will only be achieved by capturing individuality in relation to person-specific interaction and argued that there are still ways to extract regularities and generalities from individual case events. The approach taken in Chapter 9 of combining a generic coding scheme like the SAW taxonomy with network analysis may go some way to achieving this. If predictive utility of perceptual cycle–based models and methods is demonstrated, then specific training guidelines could be developed to counteract any issues uncovered in the system.

10.2.1.3 Perceptual Cycle Processes of Non-Human Agents

In Chapter 2, Schema Theory and the associated PCM were presented as a potential avenue for exploring the cognition of machines. For example, Hollnagel and Woods (1983) argued that it is reasonable to assume that explicitly designed machines possess internal models of their environments and call these the 'machines image of the operator'. They argued that this component is able to assist the machine with planning, decision-making, message formulation and message interpretation, and therefore it is akin to human schemata. Similarly, Hew (2011) argued that if there are two key functions to schema – (1) driving where and how information is gathered in the environment, and (2) adapting as a result of the gathered information – then machines demonstrate behaviour that fulfils these two functions. Additional research is required to investigate how a machine perceptual cycle might work, but this additional insight has the potential to further advance our understanding of distributed cognition. This perspective views human and non-human agents as being equally important to enable successful system functioning, and therefore the role and use of technical agents needs to be assessed giving equal attention to the artefacts that individuals utilise to fulfil the functions of a team.

10.2.2 Methodological Implications

Aside from theoretical considerations, the research presented in this book also constituted methodological developments for understanding ACDM. In turn, this has presented further avenues of research for developing, evaluating and validating the methodological factors addressed in this research.

10.2.2.1 Validate SWARM

As discussed in Chapter 7, one of the most immediate areas for future research is to continue to evaluate SWARM by assessing the reliability and validity of the prompts. The SWARM prompts are based on the theoretical perspective of the PCM. O'Hare et al. (1998) argued that the importance of theoretically driven cognitive task analysis (CTA) should not be underestimated as this has a huge potential to impact on subsequent system design and training improvements. The research in this book has studied decision-making from the perspective of the PCM, and therefore it is a logical next step to develop a method based on these theoretical principles. It is, however, only through correlating evidence with other sources that confidence in the validity and reliability of the method can be established, and this is the intention of future research. Stanton and Young (1999, p. 197) stated:

> The objective way to see whether ergonomic methods work is to assess their reliability and validity.

Chapter 7 presented an exploratory study and demonstrated the method is capable of eliciting decision-making data that provide insights into the perceptual cycle process. It would be useful for other assessments to assess the utility of SWARM over and above other data collection methods. This also needs to be continue assessment the inter- and intra-rater analyst reliability of the method (establishing this for other CTA methods like the CDM would also be best practice). As argued by Stanton et al. (2013), a method can be reliable without being valid, but it cannot be valid without being reliable. In relation to methods for extracting the knowledge of experts, Hoffman (1987) suggested that a method should be valid on three main counts: firstly, that the method discloses truths about the expert's perceptions and knowledge (internal validity); secondly, that the method is complete and covers all relevant sub-domains of knowledge (content validity); and thirdly, that there is a consensus across experts that the method is capable of measuring what it claims to measure (construct validity). Broadly speaking, the PCM prompts can be described as valid if they demonstrate that they are capable of eliciting data that describe perceptual cycle information processing. Militello and Hutton (1998) provided a rigorous assessment of a new CTA method (e.g. novices implemented the method, they undertook usability assessments with the interviewers and an 'interviewee experience' questionnaire with the interviewees); lessons of best practice can be taken from their approach to guide these future efforts. It is hoped that disseminating these prompts to like-minded researchers and practitioners will allow them to be utilised and applied in order to begin this process.

10.2.2.2 Determine Best Practice for the Elicitation of Schemata

A recurring theme throughout this research project has been the difficulty in eliciting schemata. The CDM was selected as the interview schedule for two reasons; the first being that it fulfilled the aims of the research as it was developed to elicit critical decision-making data, and secondly, previous research suggested it was a suitable method for eliciting schema-based constructs (Rafferty et al., 2012; Salmon et al., 2013). However, compared to action and world concepts, schemata were disproportionately under-represented in the data. In the data collected via communication transactions, where cognitive probes were not implemented, schema-based data were even less represented, although the richness of the action and world data was increased compared to the retrospective interview data. Future research endeavours should employ a mixed-methods approach particularly in the pursuit of perceptual cycle processes where communication transactions highlight actions and team interactions, observational techniques can assess what world information is attended to and interviews allow for retrospective reflection on the role of schema utilisation. As Neisser (1976, p. 186) argued:

> Schemata are locked inscrutably within his skull where we cannot see them.

Therefore, there will always be an element of subjectivity in the interpretation of schemata, and reliability assessments are employed to verify the credibility of these inferences, but there still exists a need for a valid and reliable method to elicit schemata in the first place, to enable them to be interpreted. The development and application of SWARM may go some way towards achieving this and will be determined by future evaluation and validation. Contemporary research has advocated the use of automatically generated network representations using context analysis software such as Leximancer™ (e.g. Rafferty et al., 2012; Salmon et al., 2013a,b; Young et al., 2015). Leximancer automates network creation by interrogating verbal transcripts to identify concepts, including relationships between concepts (Young et al., 2015). Young et al. (2015) argued that the manual construction of networks is more sensitive to differences. However, the manual analysis process is laborious and time-intensive, and therefore, more comprehensive research efforts need to investigate the advantage of using a software tool, particularly in relation to schema elicitation and representation as the automatic analysis has the advantage of objectivity compared to a manual analysis. Additionally, Yu et al. (2002) discussed the abstraction hierarchy (AH; Rasmussen, 1985) as a method that had potential to represent schemata because it can be used to demonstrate an individual's cognitive coupling with the processes they are controlling. The AH provides a hierarchical organised model of a work domain, representing both top-down and bottom-up information. This is conceptually similar to the notion of schemata (see Chapter 2). Future research needs to begin with a comprehensive method review, ideally involving the application of a number of methods to explore the elicitation of schemata so that direct comparisons can be made.

10.2.2.3 Explore the Utility of Communication Network Analysis with the EAST Methodology

Chapter 9 demonstrated that including an analysis of the specific communication transactions could provide additional insights to team processes than just using task, social and information EAST networks alone. For example, the dominant types of communication transactions exhibited by the WO, namely instruction and description, provided support for the shared leadership that existed in the SAR team and the compatible awareness achieved by the team members functioning within a distributed perceptual cycle. Artman and Garbis (1998) argued that communication and coordination are the team processes through which distributed cognition is realised; they are the transactions that allow information to pass between team members and awareness to develop. Successful team performance relies on appropriate information being communicated to the appropriate team member at the appropriate time (Stanton et al., 2009a). EAST has been established as a useful method to analyse and model distributed cognition (Walker et al., 2010; Stanton, 2014); it then stands to reason that the method would be enhanced through the addition of a communication analysis phase. Additional research is required to establish whether the descriptive power of EAST is enhanced through the consideration of communication/coordination transactions; indeed, whether benefits in additional insights outweigh the extended application time that another phase of analysis creates.

10.2.3 Practical Implications

Lipschitz et al. (2001, p. 335) have argued for 'empirical-based prescription', that is prescription should be derived from descriptive NDM models of expert performance, the goal being to improve the characteristics of a decision maker's decision process. If theoretical research is going to make any impact in the 'real' world, then it is important to consider the potential practical applications that might derive from the research. Three areas are considered, namely decision aiding and reporting, theoretical training and ecological interface design.

10.2.3.1 Decision Aids

Simpson (2001) argued that little advancement has been made in formulating techniques and strategies to teach NDM processes and proposed decision aids and decision training as means of achieving this. Similarly, Lipschitz et al. (2001) acknowledged the domain- and context-specific nature of expert knowledge and argued that NDM models that are derived from the analysis of expert knowledge should depict what information decision makers actually attend to and how they use this information. In this vein, the data and model presented in Chapter 6 have been used to develop an exemplar ACDM decision aid for technical failure (Figure 10.1). The decision aid is specific to technical failure as this was the most represented category of incident in the CDM interview data set (50%). Furthermore, there is no evidence that it is

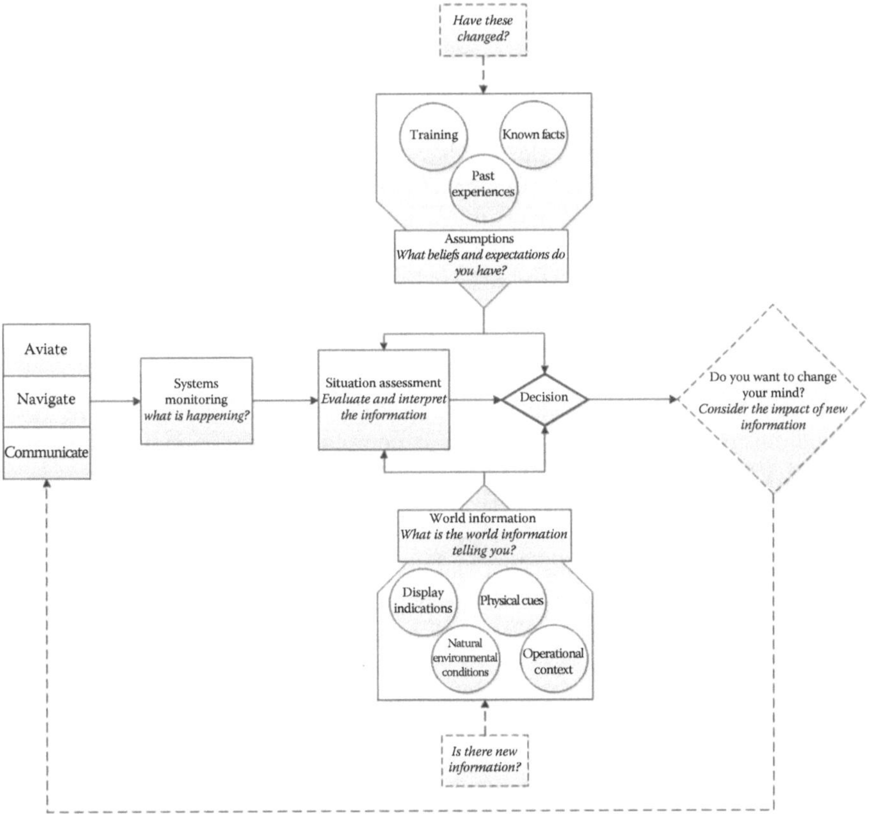

FIGURE 10.1 Exemplar ACDM decision aid for technical failure.

possible to improve all-purpose decision-making, rather specific components and skills need to be developed (Orasanu, 1994; Simpson, 2001), hence the focus on an ACDM decision aid for technical failure.

The starting point (at the far left of Figure 10.1) is the aviate–navigate–communicate strategy which is employed by all pilots when dealing with non-normal situations (to align with the triad convention, the 'location' concept has been substituted for navigate). The next step is for the decision maker to monitor the systems and ask themselves what is happening. This is followed by situation assessment, that is evaluating and interpreting the available information. The decision maker is prompted to think about the assumptions they may be bringing into this assessment process in the form of the most relevant schema types: direct past experience, trained past experience and declarative schema (known facts). They are also prompted to review the world information available to them; specifically in relation to ACDM, this includes display indications, physical cues, the natural environmental conditions and the operational context. These top-down and bottom-up sources of information also inform the decision. Finally, the decision maker is asked whether they want to change their mind by considering the impact of new information; in this instance, the process would be repeated but the decision maker is reminded to consider whether their assumptions have changed and whether there is any new information that may change their assessment of the situation or decision. This final prompt has been included to try and address the issue of humans being poor at revising probabilities based on new information (Mosier, 1991); it is intended to allow pilots to change their mind and avoid cognitive lockup or tunnel vision (O'Hare, 1992). Simpson (2001) suggested that decision aids can have a twofold benefit, firstly in helping less experienced pilots interpret the situation as a more experienced pilot would, and secondly, in helping experienced pilots make decisions under stress in a way that they would when not under stress. Additional avenues of research include the evaluation and validation of this kind of approach.

10.2.3.2 Training for Critical Decision-Making

Dealing with critical incidents are often time-limited situations, and therefore decision aids, such as the one presented in Figure 10.1, are unlikely to be used as a decision aid during a critical incident, but may be used in decision-making education. It is well acknowledged in the literature that ADM is trainable (see Li and Harris, 2008 for an overview). Further research is required to determine what the potential utility is in providing perceptual cycle–based ACDM training in a ground school environment. It can be hypothesised that with theoretical training and exposure to the perceptual cycle process of decision-making would result in its application in ACDM. Other training interventions have been employed in a similar fashion with successful outcomes. For example, O'Hare et al. (2010) found that case-based reflection training improved adherence to flight rules, with reflection improving ability at transferring and applying rules learnt from cases to new situations. A similar approach could be taken in the context of perceptual cycle-based diaries, that is participants document what they did and reflect on reasons for the decisions based on the use of schema (past experience and expectations) and world information (what was attended to).

Similarly, Deen et al. (2011) devised a programme implemented in the Oil and Gas sector which required operators to analyse incidents from both a behavioural and technical perspective. Videos of co-workers who had been involved in incidents and accidents were discussed in an integrated learning environment from the perspective of the situation, the assumptions held and resulting behaviours in a cyclical model, very similar to the PCM. Furthermore, laminated versions of the 'pyramid model' that Deen et al. (2011) utilised in their video training are also included in the 'toolbox talk' at the start of any drilling activity. This is used to remind the workers of the factors that have the potential to influence their behaviour and actions. This programme has proved highly successful at getting operators to consider the behavioural aspects of incidents, demonstrating that different realities exist in people's minds and how this can influence the way activities are approached and executed. It is envisaged that PCM-based training could be applied in a similar way and future research is necessary to discover the most successful ways of implementing this.

The team research has the potential to influence specific training content designed to enhance team functioning as a distributed cognitive unit (Grote et al., 2010). This would require a comparison between effective and less effective teams to establish the nature of differences, for example in relation to communication styles or perceptual cycle interactions (as measured by the SAW taxonomy). Salas et al. (2007) argued that team training approaches need to be grounded in a suitable theory, provide a systems perspective and guide effective behaviours and cognitions that need to be reinforced. A team training programme underpinned by the perceptual cycle would certainly fulfil these requirements, and it would be interesting to evaluate the utility of this. Furthermore, Chapter 9 touched on the potential utility of cross-training, particularly in distributed teams. There have been conflicting findings about the utility of this in the literature (see McCann et al., 2000), and therefore it is certainly an area that warrants further investigation.

10.2.3.3 Adaptive Interface Design

Research of this nature has the potential to influence the design of systems to enhance users' decision-making processes. The use of adaptive displays is gaining increasing recognition as a way of combating the potential information overload that pilots in modern aircraft cockpits face, and current interfaces do not present information in a manner that supports a full understanding of the ongoing situation (Liu et al., 2012). Studies that have evaluated the effects of adaptive displays have shown significant benefits in terms of workload reduction, improved performance and increased SA (Liu et al., 2012).

Chapter 6 demonstrated how different subtypes of the three elements of the PCM manifest at different times of ACDM and at these different phases the information requirement varies. This may have implications for the design of technological solutions to support ACDM and is exemplified with the findings from Chapter 6: in the immediate action phase, the top three most relevant concepts were display indication, trained past experience and aviating. A design solution could be envisaged whereby the pilot identifies that they are in that phase of immediate actions; they naturally continue to fly the aircraft, and their decision-making process could be assisted by the electronic provision of the relevant procedures, the de-cluttering of non-relevant display information and the enlarging of relevant display information (e.g. engine parameters for an engine-based critical incident). In the decision-making phase, the most relevant concept (after the action of making the decision) was location; therefore, this decision-making phase could be supported by the provision of a situation-relevant map detailing nearest airfields and locations that are suitable if an immediate landing is required. Similarly, Liu et al. (2012) evaluated novel adaptive displays for military pilots and found that adaptive visualisations are valuable when they highlight key aspects of the environment that impact performance. In support of this notion, Greiser et al. (2014) emphasised the use of individual pilot requirements for take-off path planning and recommended the inclusion of pilot-specific requirements and preferences in automated assistance systems.

The results of the team studies presented in Chapters 8 and 9 also have relevance for the design of adaptive displays. Stanton et al. (2009a) argued that when designing with a distributed cognitive system in mind, then displays that are designed to support team functions must be able to cater for the different roles of team members. As alluded to in Chapter 8, network-based approach could be utilised by designers to ensure they have a grasp of what information is exploited by each team member and how this information is used, that is map different information elements, both internal and external to the team members, within the networks. The interactions between the information elements of different team members also have the potential to aid designers in creating functional displays that support how different pieces of information are linked and utilised by different team members which would translate into how these are grouped and presented on an interface. For example, the PF and PNF might utilise the same display information in different ways to support their role goal, and it may be beneficial for this to be reflected in design. Due to technological advances and operational pressures, future cockpits are likely to look very different from today, for example augmented reality displays in head-up, helmet-mounted or windshield displays are likely to become the norm (Stanton et al., 2015). The highly customisable nature of these means that they have the potential to provide operationally specific or role-relevant information to enhance the acquisition and maintenance of DSA within a team.

10.3 CLOSING REMARKS

This research began back in 2010 when I started working on a European Union funded project called ALICIA (All Conditions Operations and Innovative Cockpit Architecture), looking at future flight deck technologies and their role in all conditions operations. Prior to this, my only experience of the aviation domain had been as a passenger on commercial airliners. The job of a pilot has always been glamorised, but working on the ALICIA project opened my eyes to the complex tasks they face, particularly helicopter pilots who have the added stresses of operating in difficult environments in machines with an even smaller margin of error than commercial jets. The catastrophic effects of 'pilot error' are often reported in the news. From reading the literature, it soon became apparent that the study of 'pilot error' is much more complicated than it initially appeared. It is not simply a case of back-tracking until you find a point where the human went wrong. Instead, a causal understanding of the processes involved is required. The question of *why* actions and assessments were undertaken, when operators generally do not set out to make a mistake, needs to be answered. It soon became apparent that this could only be understood if both internal and external influences to the decision maker were considered. Hence, this book set out to explore this through the perspective of the PCM. It is hoped that the work detailed in this book will promote and encourage further research into the complex issue of ACDM. The potential devastation of the consequences of system failure necessitates continued research in this area, particularly in relation to training and enhancing decision-making.

Appendix: SWARM Prompts and Procedure

A.1 REPOSITORY OF SWARM PROMPTS

Schema Categories (*Description*)	Prompt
Vicarious past experience *Experiencing something in the imagination through the description by another person (e.g. hearing a colleague recall an incident they were involved with) or documentation (e.g. reading about a certain event in an industry magazine or incident/ accident report).*	• Had you heard about or read about a similar situation happening before? • Did you draw on previously acquired knowledge of similar situations?
Direct past experience *Direct personal experience of similar events or situations in the past. This covers events experienced in live, operational contexts as opposed to those experienced through training.*	• Had the situation happened to you before? • Did you draw on previous experiences of similar situations or events? (If yes, what) • Did your previous direct experiences influence you? (If yes, how) • Was the situation comfortably within your experience? • How did your direct experience influence any expectations you may have held?
Trained past experience *Knowledge developed by direct personal experience of a specific task, event or situation, experienced within the confines of a training scenario (e.g. ground school training, simulator training or training sorties).*	• What (if any) training did you utilise? • Did your training experience influence you? (If yes, how) • How did your trained experience influence any expectations you may have held?
Declarative schema *Schema that manifests as a descriptive knowledge of facts, usually as a product of the world information available.*	• What information did you know to be true or were you certain of? (Why?) • Did any factual knowledge come to mind that assisted you? (If yes, what?) • How did information that was available to you or that you were receiving influence you? • Were you ever in doubt as to what to do? (If yes/no, why)

(*Continued*)

Schema Categories (*Description*)	Prompt
Analogical schema *Comparisons between things for the purpose of explanation and clarification. Typically, these analogies will be structural analogies of physical objects or states of affairs in the world (akin to mental map or mental model).*	• Did you create a mental image in your mind to assist you? • (If yes) What information (mental and/or physical) did you utilise to do this? • Did you imagine the consequences of your actions? • (If yes) Did this impact on or change the decision/actions that you made?
Insufficient schema *Inadequate or lacking knowledge, i.e. a schema is not developed for a certain situation.*	• Was the situation comfortably within your experience or were there parts of the situation that were unfamiliar to you? (If unfamiliar, what?) • Did you find it difficult to process or integrate information available to you because you didn't have previous experiences to draw upon? • Was there information that you weren't aware of? (If yes, what and why weren't you aware of it?) • Were there any instances of mismatches between what you were expecting to happen and what was actually happening? (If yes, why do you think this was?)

World Categories (*Description*)	Prompt
Natural environmental conditions *Natural environmental conditions (e.g. weather, light, temperature, noise).*	• What information from the natural environment did you utilise? (e.g. weather, time of day) • How and where did you get this information? • What did the environmental information tell you? • Was there additional environmental information you would have but couldn't access? (If yes, what?) • Did this environment information influence you? (If yes, how?)
Technological status *State of technological artefacts (e.g. with regards to appearance and working order)*	• What information from the technological system(s) did you utilise? • How and where did you get this information? • What did the technological information tell you? • Was there additional technological information you would have but couldn't access? (If yes, what?) • Did this technological information influence you? (If yes, how?)
Communicated information *Information available to the pilot from other people (e.g. other crew members, ATC, coastguard, etc.).*	• Did you receive information from others? (If yes, what) • How did you receive this information? • What did the communicated information tell you? • Was there additional information you would have liked to be given? (If yes, what?) • Did this communicated information influence you? (If yes, how?)
Location *Particular places or positions.*	• Where were you at this time? • What information about your location did you use to assist you?

(*Continued*)

World Categories (*Description*)	**Prompt**
Artefacts *Physical objects, including written information, symbols, diagrams or equipment.*	• What artefacts did you have available to you? (e.g. physical objects, equipment, written documents, etc.) • What artefacts (if any) did you utilise to inform your decision-making? • Did your use of artefacts change over time? (If yes, how?)
Display indications *Information elicited from the physical artefacts.*	• What were the display indications telling you? • Did information provided via the display indications assist you? (If yes, how?) • Did you change your mind about anything based on display information? (If yes, how?)
Operational context *Routine functions or activities of the organisation (e.g. search and rescue, police search, military training, etc.). This can include statements about the importance of being serviceable for the operational context or crew familiarity with the aircraft and how this effects decision-making.*	• What was the operational context? • How did the operational context influence you?
Aircraft status *Statements relating to the aircrafts integrity, performance and the general state of affairs of the aircraft at a particular time.*	• Were you concerned about the status of the aircraft? • Did the status of the aircraft influence you? (If yes, how?) • Were you ever uncertain about the reliability or relevance of the information that you had available to you?
Severity of problem *How bad (or otherwise) the critical incident is.*	• In your mind, how severe was the problem? • What information were you using to assess the severity of the problem? • Did the severity of the problem influence you? (If yes, how?)
Physical cues *External cues that provide information of conditions or states of being (e.g. noises, sounds, vibration, smells).*	• What physical cues were available to you? (e.g. sounds, smells, sights) • Did the physical cues influence you? (If yes, how?)
Absent information *Information that was missing, not present or lacking. Reasons for this may include technical faults with equipment or non-existent information.*	• Was any information missing or unavailable? (If yes, what? Why?) • Did the fact information was unavailable effect your decision-making? (If yes, how?) • Was there any other information that you would have liked to assist you? (If yes, what?)
Crew status *Statements relating to the readiness and preparedness of the crew or the current status and activities of a particular crew member.*	• What were your crew members doing at the time? • How did the crew members actions/behaviour influence your own?

(Continued)

Action Categories (*Description*)	Prompt
Aviate *Direct manipulation (handling) of flight controls in order that the aircraft can be flown and safety is maintained.*	• What physical actions did you take? • What information (mental and/or physical) did you use to assist with your physical actions? • Did you alter your actions based on new information that became available? (If yes, how?)
Navigate *The process of accurately ascertaining position and planning and following a route or desired course.*	• What navigational actions did you take? • What information (mental and/or physical) did you use to help you navigate?
Communicate *Statements relating to the explicit sharing and exchange of information or expressions used to elicit information.*	• Did you communicate with anyone? (If yes, who to/what was communicated?) • Did you share information in any other way? • Did you alter (e.g. increase or decrease) your communication actions based on new information that became available? (If yes, how?)
System management *The processes of making an input into technological systems in order that the interaction or manipulation has an explicit output.*	• What inputs did you make into the technological system? • What actions did you take that had a direct effect on the technological system? • Was there a particular time when you made more or less system management actions (If yes, why?)
System monitoring *Looking at (observing, checking) displays to gain an understanding of the situation.*	• In relation to the technological system, what were you looking at (observing or checking) during the event? • Did the information you obtained influence you in any way? (If yes, how?) • Was there a particular time when you made more or less system monitoring actions? (If yes, why?)
Environment monitoring *Observing or checking the internal or external physical environment in order to establish the current state of affairs.*	• In relation to the physical environment, what were you looking at (observing or checking)? • Did the information you obtained influence you in any way? (If yes, how?) • Was there a particular time when you made more or less environment monitoring actions? (If yes, why?)
Concurrent diagnostics *The process of determining, or attempting to determine, the cause or nature of a problem by examining the available information at the time the incident is occurring.*	• Did you attempt to diagnose the situation? (If yes, what did you decide?) • What information did you use to assist your diagnosis? • Did you seek additional information to help you? (If yes, what?) • (For critical incidents) Did you take any actions to specifically to reduce the severity of the situation? (If yes, what?)
Decision action *A conclusion or resolution that is reached after considering the available information.*	• What were the key decisions that you made? • Did you change your decision during the course of the event? (If yes, why?) • Were there other decisions that could have been chosen but weren't? (If yes, why?)

(*Continued*)

Action Categories (*Description*)	**Prompt**
Situation assessment *Actions that relate to the evaluation and interpretation of available information.*	• How did you assess the situation? • How did you evaluate and interpret the available information to you? • Based on your situation assessment, what conclusions did you draw about the event (e.g. likely outcomes)?
Non-action *Actions that were not performed either because the situation didn't warrant a particular action or because equipment faults did not allow a particular action to be performed.*	• Were there actions that you deliberately didn't perform that could have been performed? (If yes, what and why?) • Were there actions that you wanted to perform but you were unable to perform? (If yes, what and why?)
Standard operating procedure (SOP)/ checklist items *Following the prescribed procedure that ought to be routinely followed in a given situation/completing checklist items.*	• Were your actions standard and typical for the situation? (If yes/no, why/why not?) • Did you follow known standard operating procedures (or conventions or rules)? (if yes, what? If no, why not?) • Did you follow checklist procedures (memorised or physical checklists?)
Operational *Statements relating to actions that are performed for the current or intended operational situation.*	• What operational actions did you undertake?

A.2 SWARM INTERVIEW SCHEDULE

A.2.1 Schema–World–Action Research Method (SWARM)

A.2.1.1 Outline

SWARM is an interview technique to help gain an understanding of the perceptual cycle process that you, as a decision maker, engage in. There are three components to this:

1. Schema – Internally held mental templates that organise your representation of the world. Schemata are heavily influenced by past experiences and expectations (e.g. whether you have experienced the situation before and whether previously held expectations influenced the decision-making process).
2. Action – What actions were undertaken or were potentially available.
3. World – Externally available information in the world (environment), physical things, conditions or states of being and how this influenced the decision-making process.

The PCM explores the interaction that occurs between these three elements and the reciprocal, cyclical relationship that exists between them.

Research has identified different subtypes within the high-level categories, the SWARM prompts reflect these subtypes (Figure A.1).

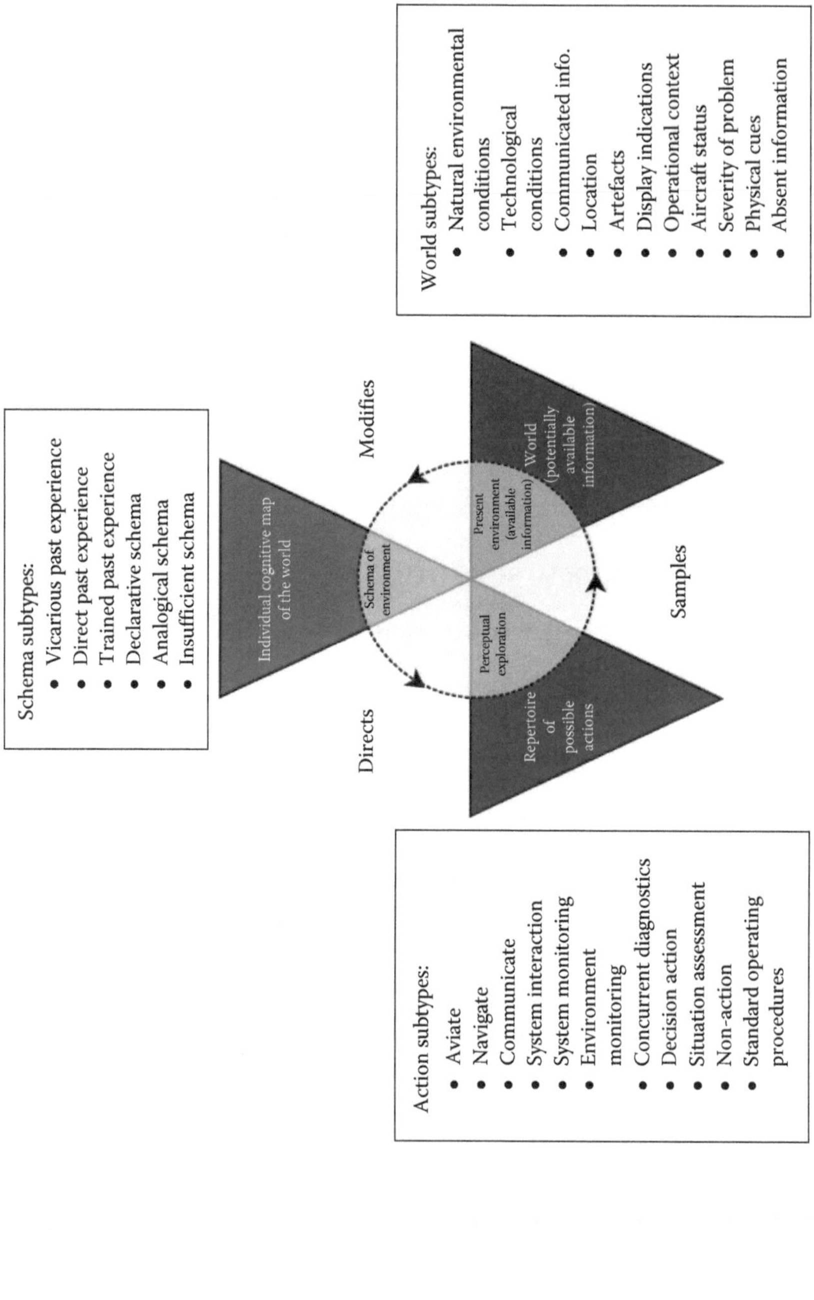

FIGURE A.1 Perceptual cycle model. (From Neisser, U., *Cognition and Reality*, W.H. Freeman & Co., San Francisco, CA, 1976.)

A.2.1.2 Demographics

Ask demographic questions relevant to the research question(s) (e.g. age, gender, date of interview, date of incident, occupation, hours of experience, role during incident, etc.).

A.2.1.3 Incident Phases

The prompt questions will be asked in relation to six phases of dealing with a critical incident. These have been identified from previous research. It may be hard to deconstruct the incident into these phases because the events are usually interlinked; it does not matter if you start talking about another phase but try to keep it separate as much as possible.

Phase 1	Pre-incident	*Set-scene and describe antecedents to the incident (i.e. weather conditions, operational context, actions undertaken)*
Phase 2	Onset of problem	*Describe the critical incident*
Phase 3	Immediate actions	*Actions to ensure immediate safety of flight based on onset of problem*
Phase 4	Decision-making	*Overall decision made for dealing with the critical incident*
Phase 5	Subsequent actions	*Additional actions taken as a result of the decision-making phase (i.e. to land immediately, land as soon as possible, resolve problem in air)*
Phase 6	Incident containment	*Conclude the account of the incident (i.e. approach, landing, shutdown, etc.)*

A.2.1.4 Scenario under Analysis

Please provide a high-level description (only taking a couple of minutes) to explain the incident that we are going to be talking about.

A.2.1.5 Phase Questioning

Each phase is associated with different PCM categories in terms of their relevance or importance in each phase.

For each phase, you will be asked to provide high-level overviews about the three PCM categories of schema, action and world. Then asked prompts in relation to some of the subtypes.

It is recommended to ask the top five subtypes and then identify other relevant ones if necessary

Questions can be skipped if they have already been answered

1: Pre-Incident	2: Onset of Problem	3: Immediate Actions	4: Decision	5: Subsequent Actions	6: Incident Containment
World: *What external information was available to you? (physical things, states of being, conditions)*					
Natural enviro. conds.	Physical cue	Display indications	Location	Communicate	Location
Operational context	Tech. conditions	System monitoring	Situation assessment	Display indication	Operational context
Location	Display indication		Aircraft status		
Are there any other relevant categories?					
Action: *What were you doing?*					
Aviate	Aviate	Aviate	Decision action	Communicate	Aviate
Standard op. procedure				Aviate	System management
				Decision action	Communicate
Are there any other relevant categories?					
Schema: *Had you experienced the situation before (past experiences), did you hold any expectations about the situation?*					
Declarative schema	Direct past exp.	Trained past exp.	Direct past exp.	Declarative schema	
	Trained past exp.	Direct past exp.			
Are there any other relevant categories?					
Additional concepts in order of relevance					
Physical cue	Natural enviro. conds.	Concurrent diagnostics	Communicate	Natural enviro. conds.	Aircraft status
Display indications	System monitoring	Decision action	Trained past exp.	Aircraft status	Declarative schema
Navigate	Concurrent diagnostic	Communicate	Absent info.	Direct past exp.	Decision action
Decision action	Absent info.	Situation assessment	Declarative schema	Insufficient schema	Non action
Situation assessment	Insufficient schema	Standard op. procedure	Aviate	Enviro. monitoring	Natural enviro. conds.
Comm'ed info.	System management	Insufficient schema	Display indication	Trained past exp.	Direct past exp.
Insufficient schema	Standard op. procedure	Non-action	Standard op. procedure	Location	Trained past exp.
Communicate	Enviro. monitoring	Artefacts	System monitoring	Operational context	Insufficient schema
Artefacts	Situation assessment	Physical cues	Operational context	Absent info.	System monitoring
Absent info.	Declarative schema	Location	Natural enviro.	System monitoring	Situation assessment

References

Air Accident Investigation Branch. 1990. Aircraft accident. Report Number: 4/90 (EW/C1095). Farnborough, UK: AAIB.

Amalberti, R. 2001. The paradoxes of almost totally safe transportation systems. *Safety Science*, 37, 109–126.

Anderson, J.P. 1995. *Cognitive Psychology and Its Implications*, 4th edn. New York: W.H. Freeman & Co.

Anderson, R.C. 1977. The notion of schema and the education enterprise: General discussions of the conference. In R.C. Anderson, R.J. Spiro and W.E. Montague (eds.), *Schooling and the Acquisition of Knowledge*. Hillsdale, NJ: Lawrence Erlbaum Associates.

Annett, J. 2002. A note of the validity and reliability of ergonomics methods. *Theoretical Issues in Ergonomics Science*, 3(2), 228–232.

Annett, J. and Stanton, N.A. 2000. Editorial: Teamwork – A problem for ergonomics? *Ergonomics*, 43(8), 1045–1051.

Artman, H. and Garbis, C. 1998. Team communication and coordination as distributed cognition. In *Proceedings of the Ninth Conference of Cognitive Ergonomics: Cognition and Cooperation*, pp. 151–156.

Axelrod, R. 1973. Schema theory: An information processing model of perception and cognition. *The American Political Science Review*, 67(4), 1248–1266.

Baber, C. 2003. *Cognition and Tool Use*. London, UK: Taylor & Francis.

Baber, C. 2006. Cognitive aspects of tool use. *Applied Ergonomics*, 37, 3–15.

Baber, C. and Stanton, N.A. 1994. Task analysis for error identification: A methodology for designing error-tolerant consumer products. *Ergonomics*, 37(11), 1923–1941.

Baber, C., Stanton, N.A., Atkinson, J., McMaster, R. and Houghton, R.J. 2013. Using social network analysis and agent-based modelling to explore information flow using common operational pictures for maritime search and rescue operations. *Ergonomics*, 56, 889–905.

Bahrick, H.P. 1984. Associations and organization in cognitive psychology: A reply to Neisser. *Journal of Experimental Psychology: General*, 113(1), 36–37.

Bainbridge, L. 1999. Verbal reports as evidence of the process operator's knowledge. *International Journal of Human-Computer Studies*, 51, 213–238.

Baker, S.P., Shanahan, D.F., Haaland, W., Brady, J.E. and Li, G. 2011. Helicopter crashes related to oil and gas operations in the Gulf of Mexico. *Aviation, Space, and Environmental Medicine*, 82(9), 885–889.

Baksteen, B. 1995. Flying is not safe. *Safety Science*, 19, 287–294.

Banks, V.A. and Stanton, N.A. 2015. Contrasting models of driver behaviour in emergencies using retrospective verbalisations and network analysis. *Ergonomics*, 58(8), 1337–1346.

Banks, V.A., Stanton, N.A. and Harvey, C. 2014. What the drivers do and do not tell you: Using verbal protocol analysis to investigate driver behaviour in emergency situations. *Ergonomics*, 53(3), 332–342.

Bartlett, F.C. 1932. *Remembering: A Study of Experimental and Social Psychology*. Cambridge, UK: Cambridge University Press.

Bellet, T., Bailly-Asuni, B., Mayenobe, P. and Banet, A. 2009. A theoretical and methodological framework for studying and modelling drivers' mental representations. *Safety Science*, 47, 1205–1221.

Bem, S.L. 1981. Gender schema theory: A cognitive account of sex typing. *Psychological Review*, 88(4), 354–364.

Bennett, S. 2001. *Human Error – By Design*? London, UK: Perpetuity Press.

Berman, B.A. and Dismukes, R.K. 2006. Pressing the approach. Accessed online [May 2015]: http://flightsafety.org/asw/dec06/asw_dec06_p28–33.pdf?dl=1.

Besnard, D., Greathead, D. and Baxter, G. 2004. When mental models go wrong. Co-occurrences in dynamic, critical systems. *International Journal of Human-Computer Studies*, 60, 117–128.

Blandford, A. and Furniss, D. 2006. DiCOT: A methodology for applying distributed cognition to the design of teamworking systems. In S.W. Gilroy and M.D. Harrison (eds.), *DSVIS 2005, LNCS* 3941, 26–38.

Blavier, A., Rouy, E., Nyssen, A.-S. and De Keyser, V. 2005. Prospective issues for error detection. *Ergonomics*, 48(7), 758–781.

Bond Aviation Group. 2015. My job at bond, winchman-winchop. Accessed online [March 2015]: http://www.bondaviationgroup.com/blog/myjobatbond-winchman-winchop.

Booth, P.A. 1991. Errors and theory in human computer interaction. In G.C. Van Der Veer, S. Bagnara and G.A.M. Kempden (eds.), *Cognitive Ergonomics, Contributions from Experimental Psychology*, pp. 69–96. Amsterdam, the Netherlands: Elsevier Science Publishers.

Bourgeon, L., Valot, C. and Navarro, C. 2013. Communication and flexibility in aircrews facing unexpected and risky situations. *The International Journal of Aviation Psychology*, 23(4), 289–305.

Bower, G.H., Black, J.B. and Turner, T.J. 1979. Scripts in memory for text. *Cognitive Psychology*, 11(2), 177–220.

Bowers, C.A., Jentsch, F., Salas, E. and Braun, C.C. 1998. Analysing communication sequences for team training needs assessment. *Human Factors*, 40(4), 672–679.

Boyatzis, R.E. 1998. *Transforming Qualitative Information: Thematic Analysis and Code Development*. Thousand Oaks, CA: Sage.

Bransford, J.D. and Johnson, M.K. 1972. Contextual prerequisites for understanding: Some investigations of comprehension and recall. *Journal of Verbal Learning and Verbal Behaviour*, 11, 717–726.

Braun, V. and Clarke, V. 2006. Using thematic analysis in psychology. *Qualitative Research in Psychology*, 3(2), 77–101.

Brewer, W.F. 1987. Schemas versus mental models in human memory. In P. Morris (ed.), *Modelling Cognition*. Chichester: Wiley, pp. 187–197.

Brewer, W.F. 2000. Bartlett's concept of the schema and its impact on theories of knowledge representation in contemporary cognitive psychology. In A. Saito (ed.), *Bartlett, Culture and Cognition*, pp. 65–89. Cambridge, UK: The Psychology Press.

Brewer, W.F. and Dupree, D.A. 1983. Use of plan schemata in the recall and recognition of goal-directed actions. *Journal of Experimental Psychology: Learning, Memory and Cognition*, 9(1), 117–129.

Brewer, W.F. and Nakamura, G.V. 1984. The nature and functions of schemas. In R.S. Wyer and T.K. Srull (eds.), *Handbook of Social Cognition*, Vol. 1, pp. 119–160. Hillsdale, NJ: Lawrence Erlbaum Associates.

Brewer, W.F. and Treyens, J.C. 1981. Role of schemata in memory for places. *Cognitive Psychology*, 13(2), 207–230.

British Helicopter Association. 2014a. Becoming a helicopter pilot. Accessed online [June 2014]: http://www.britishhelicopterassociation.org/sites/default/files/careers/pdf/Helicopter%20Pilot.pdf.

British Helicopter Association. 2014b. The future role of helicopters in public transport. Accessed online [August 2014]: http://www.britishhelicopterassociation.org/?q=about-the-bha/helicopters.

Broadbent, D.E. 1970. Frederic Charles Bartlett 1886–1969. *Biographical Memoirs of Fellows of the Royal Society*, 16, 1–13.

Burla, L., Knierim, B., Barth, J., Liewald, K., Duetz, M. and Abel, T. 2008. From text to codings. Intercoder reliability assessment in qualitative content analysis. *Nursing Research*, 57(2), 113–117.

Cacciabue, P.C. 2008. The role and challenges of Ergonomics in modern societal contexts. *Ergonomics*, 51(1), 42–48.

Carvalho, P.V.R., dos Santos, I.L. and Vidal, M.C.R. 2005. Nuclear power plant shift supervisor's decision making during microincidents. *International Journal of Industrial Ergonomics*, 35, 619–644.

Chalmers, P.A. 2003. The role of cognitive theory in human–computer interface. *Computers in Human Behaviour*, 19, 593–607.

Chimir, I.A., Abu-Dawwas, A.A. and Horney, M.A. 2005. Neisser's cycle of perception: Formal representation and practical implementation. *Journal of Computer Science*, 106–111.

Civil Aviation Authority. 1997. *CAP 667: Review of General Aviation Fatal Accidents 1985–1994*. London, UK: CAA.

Civil Aviation Authority. 1998. *CAP 681: Global Fatal Accident Review 1980–96*. London, UK: CAA.

Civil Aviation Authority. 2010. *UK Airline Statistics: 2010 – Annual*. London, UK: CAA.

Cooke, N.J., Salsa, E., Kiekel, P.A. and Bell, B. 2004. Advances in measuring team cognition. In E. Salas and S.M. Fiore (eds.), *Team Cognition: Understanding the Factors That Drive Process and Performance*. Washington, DC: American Psychological Association, pp. 83–106.

Cooper, R. and Shallice, T. 2000. Contention scheduling and the control of routine activities. *Cognitive Neuropsychology*, 17(4), 297–338.

Crandall, B. and Gretchell-Reiter, K. 1993. Critical decision method: A technique for eliciting concrete assessment indicators from the intuition of NICU nurses. *Advances in Nursing Science*, 16, 42–51.

Crandall, B., Klein, G.A. and Hoffman, R.R. 2006. *Working Minds: A Practitioner's Guide to Cognitive Task Analysis*. Cambridge, MA: MIT Press.

Deen, E., Frijters, M., Gillert, A., Swart, J. and van Wijngaarden, P. 2011. The behavioural aspects of learning from incidents. Online report from Equitans and Kessels & Smit, The Learning Company. Accessed online [February 2015]: http://www.kesselssmit.nl/files/Artikel_2011_The_behavioural_aspects_of_Learning_from_Incidents_2.pdf.

Dekker, S. 2002. The re-invention of human error. Technical Report 2002-01. Ljungbyhed, Sweden: Lund University School of Aviation.

Dekker, S. 2003. Illusions of explanation: A critical essay on error classification. *The International Journal of Aviation Psychology*, 13(2), 95–106.

Dekker, S. 2006. *The Field Guide to Understanding Human Error*. Aldershot, UK: Ashgate.

Dekker, S. 2011. What is rational about killing a patient with an overdose? Enlightenment, continental philosophy and the role of the human subject in system failure. *Ergonomics*, 54(8), 679–683.

Diehl, A. 1991. The effectiveness of training programs for preventing aircrew "Error". In *Proceedings of the Sixth International Symposium on Aviation Psychology*, pp. 640–655. Columbus, OH: The Ohio State University.

Doos, M., Backstrom, T. and Sundstrom-Frick, C. 2004. Human actions and errors in risk handling – An empirically grounded discussion of cognitive action-regulation levels. *Safety Science*, 42, 185–204.

Drivalou, S. and Marmaras, N. 2009. Supporting skill-, rule-, and knowledge-based behaviour through an ecological interface: An industry-scale application. *International Journal of Industrial Ergonomics*, 39, 947–965.

Drury, C.G. 2008. The future of ergonomics/the future of work: 45 years after Bartlett (1962). *Ergonomics*, 51(1), 14–20.

Edmonds, E.M. and Evans, S.H. 1966. Schema learning without a prototype. *Psychonomic Science*, 5, 247–248.

Elio, R. and Scharf, P.B. 1990. Modelling novice-to-expert shifts in problem solving strategy and knowledge organisation. *Cognitive Science*, 14, 579–639.

Elliott, T. 2005. Expert decision-making in naturalistic environments: A summary of research. Technical Report DSTO-GD-0429, pp. 1–61. Defence Science and Technology Organisation.

Endsley, M.R. 1995. Towards a theory of situation awareness in dynamic systems. *Human Factors*, 37(1), 32–64.

Endsley, M.R. 2000. Theoretical underpinnings of situation awareness: A critical review. In M.R. Endsley and D.J. Garland (eds.), *Situation Awareness Analysis and Measurement*. Mahwah, NJ: Lawrence Erlbaum Associates.

Endsley, M.R. and Jones, W.M. 2001. A model of inter- and intra-team situation awareness: Implications for design, training and measurement. In M. McNeese, E. Salas and M. Endsley (eds.), *New Trends in Cooperative Activities: Understanding System Dynamics in Complex Environments*, pp. 46–67. Santa Monica, CA: Human Factors and Ergonomics Society.

Ericssion, K.A. and Smith, J. 1992. *Toward a General Theory of Expertise*. Cambridge, UK: Cambridge University Press.

European Aviation Safety Agency. 2012. Annual safety review 2012. Accessed online [October 2014]: http://easa.europa.eu/system/files/dfu/EASA-Annual-Safety-Review-2012.pdf.

Evans, S.H. 1967. A brief statement of schema theory. *Psychonomic Science*, 8, 87–88.

Falzer, P.R. 2004. Cognitive Schema and naturalistic decision making in evidence-based practices. *Journal of Biomedical Informatics*, 37(2), 86–98.

Federal Aviation Authority. 2015. Online document: AC60-22 aeronautical decision making. Accessed online [April 2015]: http://www.faa.gov/regulations_policies/handbooks_manuals/aviation/pilot_handbook/media/phak%20-%20chapter%2017.pdf.

Fedota, J. and Parasuraman, R. 2009. Neuroergonomics and human error. *Theoretical Issues in Ergonomics Science*, 11(5), 1–20.

Flach, J.M., Dekker, S. and Stappers, P.J. 2008. Playing twenty questions with nature (the surprise version): Reflections on the dynamics of experience. *Theoretical Issues in Ergonomics Science*, 9(2), 125–154.

Flanagan, J.C. 1954. The critical incident technique. *Psychological Bulletin*, 51(4), 327–358.

Fleishman, E.A. and Quaintance, M.K. 1984. *Taxonomies of Human Performance: The Description of Human Tasks*. Orlando, FL: Academic Press.

Fletcher, G., Flin, R., McGeorge, P., Glavin, R., Maran, N. and Patey, R. 2004. Rating non-technical skills: Developing a behavioural marker system for use in anaesthesia. *Cognition, Technology and Work*, 6, 165–171.

Flyvbjerg, B. 2011. Case study. In N.K. Denzin and Y.S. Lincoln (eds.), *The Sage Handbook of Qualitative Research*, 4th edn., pp. 301–316. Thousand Oaks, CA: Sage.

Goode, N., Salmon, P.M., Lenne, M.G. and Hillard, P. 2014. Systems thinking applied to freight handling operations. *Accident Analysis and Prevention*, 68, 181–191.

Grasser, A.C. and Nakamura, G.V. 1982. The impact of schemata on comprehension and memory. In G.H. Bower (ed.), *The Psychology of Learning and Motivation: Advances in Research and Theory*, Vol. 16. New York: Academic Press.

Greiser, S., Wolfman, J. and Schieben, A. 2014. The role of the helicopter pilot in terms of relevant requirements for helicopter takeoff. *The International Journal of Aviation Psychology*, 24(3), 172–189.

Griffin, T.G.C., Young, M.S. and Stanton, N.A. 2010. Investigating accident causation through information network modelling. *Ergonomics*, 53(2), 198–210.

Groeger, J.A. 1997. *Memory and Remembering: Everyday Memory in Context*. London, UK: Longman.

Grote, G., Kolbe, M., Zala-Mezo, E., Bienfeld-Seall, N. and Kunzle, B. 2010. Adaptive coordination and heedfulness make better cockpit crews. *Ergonomics*, 53(2), 211–228.

Hall, J.G. and Silva, A. 2008. A conceptual model for the analysis of mishaps in human operated safety-critical systems. *Safety Science*, 46, 22–37.

Hancock, P.A. and Diaz, D.D. 2002. Ergonomics as a foundation for a science of purpose. *Theoretical Issues in Ergonomics Science*, 3(2), 115–123.

Hancock, P.A., Hancock, G.M. and Warm, J.S. 2009. Individuation: The N = 1 revolution. *Theoretical Issues in Ergonomics Science*, 10(5), 481–488.

Hannigan, S.L. and Tippens Reunitz, M. 2001. A demonstration and comparison of two types of inference-based memory errors. *Journal of Experimental Psychology: Learning, Memory and Cognition*, 27(4), 931–940.

Harris, D. and Li, W.-C. 2010. An extension of the human factors analysis and classification system for use in open systems. *Theoretical Issues in Ergonomics Science*, 12(2), 1–21.

Harris, D. and Stanton, N.A. 2010. Editorial: Aviation as a system of systems: Preface to the special issue of human factors in aviation. *Ergonomics*, 53(2), 145–148.

Henley, I., Anderson, P. and Wiggins, M. 1999. Integrating human factors education in general aviation: Issues and teaching strategies. In D. O'Hare (ed.), *Human Performance in General Aviation*, pp. 89–117. Aldershot, UK: Ashgate.

Hew, P. 2011. Reconciling situation awareness in humans versus machines via mode awareness and schemata. *Theoretical Issues in Ergonomics Science*, 14(4), 330–351.

Hobbs, A. and Williamson, A. 2002. Skills, rules and knowledge in aircraft maintenance: Errors in context. *Ergonomics*, 45(4), 209–308.

Hoffman, R.R. 1987. The problem of extracting the knowledge of experts from the perspective of experimental psychology. *AI Magazine*, 8(2), 53–67.

Hoffman, R.R., Crandall, B. and Shadbolt, N. 1998. Use of the critical decision method to elicit expert knowledge: A case study in the methodology of cognitive task analysis. *Human Factors*, 40(2), 254–276.

Hole, G. 2007. *The Psychology of Driving*. Hillsdale, NJ: Lawrence Erlbaum Associates.

Holland, D. 1992. Chapter 4: The woman who climbed up the house: Some limitations of schema theory. In T. Schwartz, G.M. White and C.A. Lutz (eds.), *New Directions in Psychological Anthropology*. Cambridge, UK: Cambridge University Press.

Hollnagel, E. 1993. *Human Reliability Analysis: Context and Control*. London, UK: Academic Press.

Hollnagel, E. 2001. Extended cognition and the future of Ergonomics. *Theoretical Issues in Ergonomics Science*, 2(3), 309–315.

Hollnagel, E., Kaarstad, M. and Lee, H.C. 1999. Error mode prediction. *Ergonomics*, 42(11), 1457–1471.

Hollnagel, E. and Woods D.D. 1983. Cognitive systems engineering: New wine in new bottles. *International Journal of Man-Machine Studies*, 18, 583–600.

Hollnagel, E. and Woods, D.D. 2005. *Joint Cognitive Systems: Foundations of Cognitive Systems Engineering*. Boca Raton, FL: CRC Press.

Hollywell, P.D. 1996. Incorporating human dependent failures in risk assessments to improve estimates of actual risk. *Safety Science*, 22, 177–194.

Houghton, R.T., Baber, C., McMaster, R., Stanton, N.A., Salmon, P.M., Stewart, R. and Walker, G. 2006. Command and control in emergency services operations: A social network analysis. *Ergonomics*, 49(12–13), 1204–1225.

Hutchins, E. 1995a. *Cognition in the Wild*. Cambridge, MA: The MIT Press.

Hutchins, E. 1995b. How a cockpit remembers its speed. *Cognitive Science*, 19, 265–288.

Hutchins, E. 2000. Distributed cognition. *International Encyclopaedia of the Social and Behavioural Sciences.* Accessed online [November 2013]: http://hip.cntb.webfactional.com/hipsite/uploads/2014/07/Hutchins_DistributedCognition.pdf.

Jarvis, S. and Harris, D. 2010. Development of a bespoke human factors taxonomy for gliding accident analysis and its revelations about highly inexperienced UK glider pilots. *Ergonomics*, 53(2), 294–303.

Jenkins, D., Salmon, P.M., Stanton, N.A., Walker, G.H. and Rafferty, L. 2011. What could they have been thinking? How sociotechnical system design influences cognition: A case study of the Stockwell shooting. *Ergonomics*, 54(2), 103–119.

Jentsch, F. and C. Bowers. 2005. Team communication analysis. In N.A. Stanton, A. Hedge, K. Brookhuis, E. Salas and H. Hendrick (eds.), *Handbook of Human Factors and Ergonomics Methods*, pp. 50.1–50.5. Boca Raton, FL: CRC Press.

Johnson, C.W. 1997. The epistemics of accidents. *International Journal of Human-Computer Studies*, 47, 659–688.

Johnson, C.W., McCarthy, A.C. and Wright, P.C. 1995. Using a formal language to support natural language in accident reports. *Ergonomics*, 38(6), 1264–1282.

Johnson-Laird, P. 1983a. *Mental Models.* Cambridge, MA: Harvard University Press.

Johnson-Laird, P.N. 1983b. *Mental Models: Towards a Cognitive Science of Language, Inference and Consciousness.* Cambridge, UK: Cambridge University Press.

Johnston-Conover, P. and Feldman, S. 1991. Where is the schema? Critiques. *The American Political Science Review*, 85(4), 1364–1369.

Kanaki, B.G. 1995. A training perspective: Enhancing team performance through effective communication. In B.G. Kanaki and O.V. Prinzo (eds.), *Proceedings of the Methods and Metrics of Voice Communications Workshop*, pp. 39–42. San Antonio, TX: Federal Aviation Administration.

Kerlinger, F.N. and Lee, H.B. 2000. *Foundations of Behavioural Research*, 4th edn. New York: Harcourt College Publishers.

Kirlik, A. and Strauss, R. 2006. Situation awareness as judgement. I: Statistical modelling and quantitative measurement. *International Journal of Industrial Ergonomics*, 36, 463–474.

Klayman, J. and Ha, Y.-W. 1987. Confirmation, disconfirmation, and information in hypothesis testing. *Psychological Review*, 94, 211–228.

Klein, G. 1998. *Sources of Power. How People Make Decisions.* Cambridge, MA: MIT Press.

Klein, G. 2008. Naturalistic decision making. *Human Factors*, 50(3), 456–460.

Klein, G.A. 1993. Twenty questions, suggestions for research in naturalistic decision making. In G.A. Klein, J. Orasanu, R. Calder-wood and C.E. Zsambok (eds.), *Decision Making in Action: Models and Methods*, pp. 389–403. Norwood, NJ: Ablex.

Klein, G.A. and Armstrong, A. 2005. Critical decision method. In N.A. Stanton, A. Hedge, K. Brookhuis, E. Salas and H. Hendrick (eds.), *Handbook of Human Factors and Ergonomics Methods*, pp. 35.1–35.8. Boca Raton, FL: CRC Press.

Klein, G.A., Calderwood, R. and Clinton-Cirocco, A. 1986. Rapid decision making on the fire ground. In *30th Annual Human Factors Society Meeting*, Dayton, OH.

Klein, G.A., Calderwood, R. and Macgregor, D. 1989. Critical decision method for eliciting knowledge. *IEEE Transactions on Systems, Man and Cybernetics*, 19(3), 462–472.

Kontogiannis, T. and Malakis, S. 2009. A proactive approach to human error detection and identification in aviation and air traffic control. *Safety Science*, 47, 693–706.

Kuklinski, J.H., Luskin, R.C. and Bolland, J. 1991. Where is the schema? Critiques. *The American Political Science Review*, 85(4), 1357–1380.

Langan-Fox, J., Canty, J.M. and Sankey, M.J. 2009. Human-automation teams and adaptable control for future air traffic management. *International Journal of Industrial Ergonomics*, 39, 894–903.

Lenne, M.G., Ashby, K. and Fitzharris, M. 2008. Analysis of general aviation crashed in Australia using the human factors analysis and classification system. *The International Journal of Aviation Psychology*, 18(4), 340–352.

Levi, D. and Slem, C. 1995. Team work in research and development organisations: The characteristics of successful teams. *International Journal of Industrial Ergonomics*, 16, 29–42.

Leximancer. 2011. Leximancer user manual: From words to meaning to insight, v3.5. www.leximancer.com.

Li, W.-C. and Harris, D. 2001. The evaluation of the effect of a short aeronautical decision-making training program for military pilots. *The International Journal of Aviation Psychology*, 18(2), 135–152.

Lieberman, D.A. 2012. *Human Learning and Memory.* Cambridge, MA: Cambridge University Press.

Lipshitz, R., Klein, G., Orasanu, J. and Salas, E. 2001. Focus article: Taking stock of naturalistic decision making. *Journal of Behaviour Decision Making*, 14, 331–352.

Lipshitz, R. and Shaul, S.B. 1997. Schemata and mental models in recognition-primed decision making. In C.E. Zsambok and G. Klein (eds.), *Naturalistic Decision Making*, pp. 60–72. Mahwah, NJ: Lawrence Erlbaum Associates.

Liu, D., Guarino, S.L., Roth, E., Harper, K. and Vincenzi, D. 2012. Effect of novel adaptive displays on pilot performance and workload. *The International Journal of Aviation Psychology*, 22(3), 242–265.

Lodge, M. and McGraw, K.M. 1991. Where is the schema? Critiques. *The American Political Science Review*, 85(4), 1357–1364.

Lombard, M., Snyder-Duch, J. and Bracken, C.C. 2002. Content analysis in mass communication. Assessment and reporting of intercoder reliability. *Human Communication Research*, 28(4), 587–604.

Magazzu, D., Comelli, M. and Marinoni, A. 2006. Are car drivers holding a motorcycle licence less responsible for motorcycle-car crash occurrence? A non-parametric approach. *Accident Analysis and Prevention*, 38(2), 365–370.

Malaterre, G. 1986. Errors analysis and in-depth accident studies. *Ergonomics*, 33, 1403–1421.

Mandler, J.M. 1984. *Stories Scripts and Scenes: Aspects of Schema Theory.* Hillsdale, NJ: Lawrence Erlbaum Associates.

Marques, J. and McCall, C. 2005. The application of inter-rater reliability as a solidification instrument in a phenomenological study. *The Qualitative Report*, 10(3), 439–462.

Marshall, S.P. 1995. *Schemas in Problem Solving.* Cambridge, UK: Cambridge University Press.

Maurino, D.E.M. 2000. Human factors and aviation safety: What the industry has, what the industry needs. *Ergonomics*, 43(7), 952–959.

McCann, C., Baranski, J.V., Thompson, M.M. and Pigeau, R.A. 2000. On the utility of experiential cross-training for team decision-making under time stress. *Ergonomics*, 43(8), 1095–1110.

McFadden, K.L. and Towell, E.R. 1999. Aviation human factors: A framework for the new millennium. *Journal of Air Transport Management*, 5, 177–184.

Militello, L.G. and Hutton, R.J.B. 1998. Applied cognitive task analysis (ACTA): A practitioner's toolkit for understanding cognitive task demands. *Ergonomics*, 41(11), 1618–1641.

Miller, A.H. 1991. Where is the schema? Critiques. *The American Political Science Review*, 85(4), 1369–1376.

Minsky, M. 1975. A framework for representing knowledge. In P. Winston (ed.), *The Psychology of Computer Vision*, pp. 211–277. New York: McGraw-Hill.

Mitchell, R.J., Williamson, A.M., Molesworth, B. and Chung, A.Z.Q. 2014. A review of the use of human factors classification frameworks that identify causal factors for adverse events in the hospital setting. *Ergonomics*, 57(10), 1443–1472.

Moir, S., Paquet, V., Punnett, L., Buchholz, B. and Wegman, D. 2003. Making sense of highway construction: A taxanomic framework for ergonomic exposure assessment and intervention research. *Applied Occupational and Environmental Hygiene*, 18(4), 256–267.

Mosier, K.L. 1991. Expert decision making strategies. In R.S. Jensen (ed.), *Proceedings of the Sixth International Symposium of Aviation Psychology*. Columbus, OH: Ohio State University.

Moray, N. 1990. A lattice theory approach to the structure of mental models. *Philosophical Transactions of the Royal Society*, 327, 577–583.

Moray, N. 1996. A taxonomy and theory of mental models. In *Proceedings of the Human Factors and Ergonomics Society Annual Meeting*, Vol. 40, pp. 164–168.

Moray, N. 1999. Mental models in theory and practice. In D. Gopher and A. Koriat (eds.), *Attention and Performance XVII: Cognitive Regulation of Performance: Interaction of Theory and Application*, pp. 223–258. Cambridge, MA: MIT Press.

Morris, C.H. and Leung, Y.K. 2006. Pilot mental workload: How well do pilots really perform? *Ergonomics*, 49(15), 1581–1596.

Morrison, J.G., Kelly, R.T., Moore, R.A. and Hutchins, S.G. 1997. Tactical decision making under stress (TADMUS) decision support system. In *Proceedings of the 1997 National Symposium on Sensor and Data Fusion*. Lexington, MA: MIT Lincoln Laboratory.

Nascimiento, F., Majumdar, A. and Ochieng, W.Y. 2014. Helicopter accident analysis. *The Journal of Navigation*, 67, 145–161.

National Transport Safety Board. 2011. Annual review of aircraft accident data, U.S. air carrier operations: Calendar year 2011. Report Number: NTSB/ARA-14/01. Washington, DC: NTSB.

Neissen, C., Eyeferth, K. and Bierwagen, T. 1999. Modelling cognitive processes of experienced air traffic controllers. *Ergonomics*, 42(11), 1507–1520.

Neisser, U. 1967. *Cognitive Psychology*. New York: Appleton Century Crofts.

Neisser, U. 1976. *Cognition and Reality*. San Francisco, CA: W.H. Freeman & Co.

Norman, D.A. 1981. Categorization of action slips. *Psychological Review*, 88(1), 1–15.

Norman, D.A. and Shallice, T. 1986. Attention to action. In R.J. Davidson, G.E. Schwartz and Shapiro, D. (eds.), *Consciousness and Self-Regulation*, pp. 1–18. New York: Plenum Press.

O'Hare, D. 1992. The 'Artful' decision maker: A framework model for aeronautical decision making. *The International Journal of Aviation Psychology*, 2(3), 175–191.

O'Hare, D. 2000. The 'Wheel of Misfortune': A taxonomic approach to human factors in accident investigation and analysis in aviation and other complex systems. *Ergonomics*, 43(12), 2001–2009.

O'Hare, D., Mullen, N. and Arnold, A. 2010. Enhancing aeronautical decision making through case-based reflection. *The International Journal of Aviation Psychology*, 20(1), 48–58.

O'Hare, D., Wiggins, M., Batt, R. and Morrison, D. 1994. Cognitive failure analysis for aircraft accident investigation. *Ergonomics*, 37(11), 1855–1869.

O'Hare, D., Wiggins, M., Williams, A. and Wong, W. 1998. Cognitive task analyses for decision centred design and training. *Ergonomics*, 41(11), 1698–1718.

Oldfield, R.C. 1972. Frederic Charles Bartlett: 1886–1969. *American Journal of Psychology*, 85, 132–140.

Orasanu, J. 1994. Shared problem models and flight crew performance. In N. Johnston, N. McDonald and R. Fuller (eds.), *Aviation Psychology in Practice*, pp. 2–30. Aldershot, UK: Ashgate.

Orasanu, J. and Martin, L. 1998. Errors in aviation decision making: A factor in accidents and incidents. In *Proceedings of the Second Workshop on Human Error, Safety and System Development*, Seattle, WA.

Paris, C., Hall Johnston, J. and Reeves, D. 2000a. A schema-based approach to measuring team decision making in a navy combat information centre. In C. McCann and R. Pigeau (eds.), *The Human in Command. Exploring the Modern Military Experience.* New York: Kluwer Academic Publishers.

Paris, C.R., Salas, E. and Cannon-Bowers, J.A. 2000b. Teamwork in multi-person systems: A review and analysis. *Ergonomics*, 43(8), 1052–1075.

Patrick, J. and Morgan, P.L. 2010. Approaches to understanding, analysing and developing situation awareness. *Theoretical Issues in Ergonomics Science*, 11(1), 41–57.

Patton, M.Q. 1990. *Qualitative Evaluation and Research Methods*, 2nd edn. Newbury Park, CA: Sage.

Piaget, J. 1952. *The Origins of Intelligence in Children.* New York: International Universities Press.

Pidgeon, N. and O'Leary, M. 2000. Man-made disasters: Why technology and organizations (sometimes) fail. *Safety Science*, 34, 15–30.

Piegorsch, K.M., Watkins, K.W., Piegorsch, W.W., Reininger, B., Corwin, S.J. and Valois, R.F. 2006. Ergonomic decision making: A conceptual framework for experienced practitioners from backgrounds in industrial engineering and physical therapy. *Applied Ergonomics*, 37, 587–598.

Plant, K.L. and Stanton, N.A. 2011. A critical incident in the cockpit: Analysis of a critical incident interview using the Leximancer™ tool. In *Proceedings of the Third International Conference of the European Aerospace Societies*, pp. 486–492. Venice, FL: CEAS.

Plant, K.L. and Stanton, N.A. 2012a. Why did the pilots shut down the wrong engine? Explaining errors in context using Schema Theory and the Perceptual Cycle Model. *Safety Science*, 50, 300–315.

Plant, K.L. and Stanton, N.A. 2012b. 'I did something against all regulations': Decision making in critical incidents. In S. Landry (ed.), *Advances in Human Aspects of Aviation.* Boca Raton, FL: CRC Press.

Plant, K.L. and Stanton, N.A. 2013a. The explanatory power of schema theory: Theoretical foundations and future applications in ergonomics. *Ergonomics*, 51(6), 1–15.

Plant, K.L. and Stanton, N.A. 2013b. What's on your mind? Using the Perceptual Cycle Model and Critical Decision Method to understand the decision-making process in the cockpit. *Ergonomics*, 56(8), 1232–1250.

Plant, K.L. and Stanton, N.A. 2014a. The process of processing: Exploring the validity of Neisser's perceptual cycle with accounts from critical decision-making in the cockpit. *Ergonomics*, 58, 909–923.

Plant, K.L. and Stanton, N.A. 2014b. All for one and one for all: Representing teams as a collection of individuals and an individual collective using a network perceptual cycle approach. *International Journal of Industrial Ergonomics*, 44, 777–792.

Ponomarenko, V.A. 2004. The significance of theoretical concepts in activity theory for applied research in aviation. *Theoretical Issues in Ergonomics Science*, 5(4), 297–312.

Posner, M.I. and Keele, S.W. 1968. On the genesis of abstract ideas. *Journal of Experimental Psychology*, 77, 353–363.

Posner, M.I. and Keele, S.W. 1970. Retention of abstract ideas. *Journal of Experimental Psychology*, 83, 304–308.

Rafferty, L.A., Stanton, N.A. and Walker, G.H. 2010. The famous five factors in teamwork: A case study of fratricide. *Ergonomics*, 53(10), 1187–1204.

Rafferty, L.A., Stanton, N.A. and Walker, G.H. 2012. *The Human Factors of Fratricide.* Aldershot, UK: Ashgate.

Rafferty, L.A., Stanton, N.A. and Walker, G.H. 2013. Great expectations: A thematic analysis of situation awareness in fratricide. *Safety Science*, 56, 63–71.

Rasmussen, J. 1974. The human data processor as a system component: Bits and pieces of a model. Report No. RisØ-M-1722. Roskilde, Denmark: Danish Atomic Energy Commission.

Rasmussen, J. 1983. Skills, rules, and knowledge; signals, signs, and symbols, and other distinctions in human performance models. *IEEE Transactions on Systems, Man, and Cybernetics*, SMC-13, 257–266.

Rasmussen, J. 1985. The role of hierarchal knowledge representation in decision making and system management. *IEEE Transactions on Systems, Man and Cybernetics*, 15, 234–243.

Rasmussen, J. 1990. Human error and the problem of causality in analysis of accidents. *Philosophical Transactions of the Royal Society of London*, 327, 449–462.

Rasmussen, J. 1993. Diagnostic reasoning in action. *IEEE Transactions on Systems, Man and Cybernetics*, 23, 981–992.

Reason, J. 1988. Modelling the basic error tendencies of human operators. *Reliability Engineering and System Safety*, 22, 137–153.

Reason, J. 1990. *Human Error.* Cambridge, UK: Cambridge University Press.

Reason, J. 2000. Human errors: Models and management. *British Medical Journal*, 320, 769–770.

Revell, K. and Stanton, N.A. 2012. Models of models: Filtering and bias rings in depiction of knowledge. *Ergonomics*, 55(9), 1073–1092.

Richardson, M. and Ball, L.J. 2009. Internal representations, external representations and ergonomics: Towards a theoretical integration. *Theoretical Issues in Ergonomics Science*, 10(4), 335–376.

Robson, C. 2002. *Real World Research. A Resource for Social Scientists and Practitioner-Researchers.* Oxford, UK: Blackwell Publishing.

Rodrigues de Carvalho, P.V., Gomes, J.O., Huber, G.J. and Vidal, M.C. 2009. Normal people working in normal organizations with normal equipment: System safety and cognition in a mid-air collision. *Applied Ergonomics*, 40, 325–340.

Rumelhart, D.E. and Ortony, A. 1977. The representation of knowledge in memory. In R.C. Anderson, R.J. Sapiro and W.E. Montague (eds.), *Schooling and the Acquisition of Knowledge*, pp. 99–135. Hillsdale, NJ: Lawrence Erlbaum Associates.

Rybing, J., Nilsson, H., Jonson, C.-O. and Bang, M. 2015. Studying distributed cognition of simulation-based team training with DiCoT. *Ergonomics*, 4, 1–12.

Sadoski, M., Paivio, A. and Goetz, E.T. 1991. A critique of schema theory in reading and a dual coding theory alternative. *Reading Research Quarterly*, 26(4), 463–484.

Saito, A. 2000. *Bartlett, Culture and Cognition*. Cambridge, UK: The Psychology Press.

Salas, E. 2005. Team methods. In N.A. Stanton, A. Hedge, K. Brookhuis, E. Salas and H. Hendrick (eds.), *Handbook of Human Factors and Ergonomics Methods*, pp. 43.1–43.3. Boca Raton, FL: CRC Press.

Salas, E., Guthrie, J. and Burke, S. 2007. Why training team decision making is not as easy as you think: Guiding principles and needs. In M. Cook, J. Noyes and Y. Masakawski (eds.), *Decision Making in Complex Environments*, pp. 225–232. Aldershot, UK: Ashgate.

Salmon, P.M., Cornelissen, M. and Trotter, M.J. 2012. Systems-based accident analysis methods: A comparison of Accimap, HFACS, and STAMP. *Safety Science*, 50, 1158–1170.

Salmon, P.M., Lenne, M.G., Walker, G.H., Stanton, N.A. and Filtness, A. 2014a. Exploring schema-driven differences in situation awareness between road users: An on-road study of driver, cyclist and motorcyclist situation awareness. *Ergonomics*, 57(2), 191–209.

Salmon, P.M., Lenne, M.G., Walker, G.H., Stanton, N.A. and Filtness, A. 2014b. Using the Event Analysis of Systemic Teamwork (EAST) to explore conflicts between different road user groups when making right hand turns at urban intersections. *Ergonomics*, 57(11), 1628–1642.

Salmon, P.M., Read, G.J.M., Stanton, N.A. and Lenne, M.G. 2013a. The crash at Kerang: Investigating systemic and psychological factors leading to unintentional non-compliance at rail level crossings. *Accident Analysis and Prevention*, 50, 1278–1288.

Salmon, P.M., Stanton, N.A., Walker, G.H., Baber, C., Jenkins, D.P., McMaster, R. and Young, M.S. 2008a. What is really going on? Review of situation awareness models for individuals and teams. *Theoretical Issues in Ergonomics Science*, 9(4), 297–323.

Salmon, P.M., Stanton, N.A., Walker, G.H. and Jenkins, D.P. 2009a. *Distributed Situation Awareness: Theory Measurement and Application to Teamwork.* Aldershot, UK: Ashgate.

Salmon, P.M., Stanton, N.A., Walker, G.H., Jenkins, D.P., Baber, C. and McMaster, R. 2008b. Representing situation awareness in collaborative systems: A case study in the energy distribution domain. *Ergonomics*, 51(3), 367–384.

Salmon, P.M., Stanton, N.A., Walker, G.H., Jenkins, D.P., Ladva, D., Rafferty, L. and Young, M. 2009b. Measuring situation awareness in complex systems: Comparison of measures study. *International Journal of Industrial Ergonomics*, 39, 490–500.

Salmon, P.M., Walker, G.H. and Stanton, N.A. 2015. Broken components versus broken systems: Why it is systems not people that lose situation awareness. *Cognition, Technology, and Work*, 17, 179–183.

Salmon, P.M., Young, K.L. and Cornelissen, M. 2013b. Compatible cognition amongst road users: The compatibility of driver, motorcyclist, and cyclist situation awareness. *Safety Science*, 56, 6–17.

Sarter, N. and Alexander, J. 2000. Error types and related error detection mechanisms in the aviation domain: An analysis of aviation safety reporting system incident reports. *The International Journal of Aviation Psychology*, 10(2), 189–206.

Schank, R.C. 1982. *Dynamic Memory: A Theory of Reminding and Learning in Computers and People.* Cambridge, UK: Cambridge University Press.

Schank, R.C. and Abelson, R. 1977. *Scripts, Plans, Goals and Understanding: An Enquiry into Human Knowledge Structures.* Hillsdale, NJ: Lawrence Erlbaum Associates.

Schmidt, R.A. 1975. A schema theory of discrete motor skill learning. *Psychological Review*, 8(4), 225–260.

Schmidt, R.A. 2003. Motor schema theory after 27 years: Reflections and implications for a new theory. *Research Quarterly for Exercise and Sport*, 74(4), 366–375.

Schutte, P.C. and Trujillo, A.C. 1996. Flight crew task management in non-normal situations. In *Proceedings of the 40th Annual Meeting of the Human Factors and Ergonomics Society*, pp. 244–248. Santa Monica, CA: HFES.

Sexton, J.B. and Helmreich, R.L. 2000. Analysing cockpit communications: The Links between language, performance, error, and workload. *Journal of Human Performance in Extreme Environments*, 5(1), 62–68.

Shappell, S. and Wiegmann, D. 2000. The human factors analysis and classification system (HFACS). Report Number DOT/FAA/AM-00/7. Washington, DC: FAA.

Shappell, S. and Wiegmann, D. 2009. A methodology for assessing safety programs targeting human error in aviation. *The International Journal of Aviation Psychology*, 19(3), 252–269.

Simon, H.A. 1957. *Models of Man: Social and Rational; Mathematical Essays on Rational Human Behaviour in a Social Setting.* New York: Wiley.

Simon, H.A. 1959. Theories of decision making in economics and behavioural science. *The American Economics Review*, 49(3), 253–283.

Simpson, P.A. 2001. Naturalistic decision making in aviation environments. Report No. DSTO-GD-0279. Bend, Victoria, Australia: DSTO Aeronautical and Maritime Research Laboratory.

Singletary, M.W. 1993. *Mass Communication Research: Contemporary Methods and Applications.* Boston, MA: Addison-Wesley.

Smith, D.E. and Marshall, S. 1997. Applying hybrid models of cognition in decision aids. In C.E. Zsambok and G.A. Klien (eds.), *Naturalistic Decision Making*, pp. 331–341. Mahwah, NJ: Lawrence Erlbaum Associates.

Smith, K. and Hancock, P.A. 1995. Situation awareness is adaptive, externally directed consciousness. *Human Factors*, 37(1), 137–148.

Smith-Jackson, T.L. and Wogalter, M.S. 2007. Application of a mental models approach to MSDS design. *Theoretical Issues in Ergonomics Science*, 8(4), 303–319.

Sorensen, L.J. and Stanton, N.A. 2011. Is SA shared or distributed in teamwork? An exploratory study in an intelligence analysis task. *International Journal of Industrial Ergonomics*, 41(6), 677–687.

Sorensen, L.J. and Stanton, N.A. 2013. Y is best: How distributed situation awareness is mediated by organisational structure and correlated with task success. *Safety Science*, 56, 72–79.

Stanton, N.A. 2002. Editorial: Developing and validating theory in ergonomics science. *Theoretical Issues in Ergonomics Science*, 3(2), 111–114.

Stanton, N.A. 2005. Behaviour and cognitive methods. In N.A. Stanton, A. Hedge, K. Brookhuis, E. Salas and H. Hendrick (eds.), *Handbook of Human Factors and Ergonomics Methods*, pp. 27.1–27.8. Boca Raton, FL: CRC Press.

Stanton, N.A. 2014. Representing distributed cognition in complex systems: How a submarine returns to periscope depth. *Ergonomics*, 57(3), 403–418.

Stanton, N.A. and Baber, C. 1996. A systems approach to human error identification. *Safety Science*, 22, 215–228.

Stanton, N.A. and Baber, C. 2005. Validating task analysis for error identification: Reliability and validity of a human error prediction technique. *Ergonomics*, 48(9), 1097–1113.

Stanton, N.A., Baber, C. and Harris, D. 2008. *Modelling Command and Control*. Aldershot, UK: Ashgate.

Stanton, N.A., Chambers, P.R.G. and Piggott, J. 2001. Situation awareness and safety. *Safety Science*, 39, 189–204.

Stanton, N.A., Plant, K.L., Roberts, A.P. and Harvey, C. 2016. Extending helicopter operations to meet future integrated transport needs. *Applied Ergonomics*, 53, 364–373.

Stanton, N.A., Rafferty, L.A., Salmon, P.M., Revell, K.M., McMaster, R., Caird-Daley, A. and Cooper-Chapman, C. 2010a. Distributed decision making in multihelicopter teams: Case study of mission planning and execution from a non-combatant evacuation operation training scenario. *Journal of Cognitive Engineering and Decision Making*, 4(4), 328–353.

Stanton, N.A. and Salmon, P.M. 2009. Human error taxonomies applied to driving: Generic driver error taxonomy and its implications for intelligent transport systems. *Safety Science*, 47, 227–237.

Stanton, N.A., Salmon, P.M., Rafferty, L.A., Walker, G.H., Baber, C. and Jenkins, D. 2013. *Human Factor Methods. A Practical Guide for Engineering and Design*, 2nd edn. Aldershot, UK: Ashgate.

Stanton, N.A., Salmon, P.M. and Walker, G.H. 2015. Let the reader decide: A paradigm shift for situation awareness in sociotechnical systems. *Journal of Cognitive Engineering and Decision Making*, 9(1), 44–50.

Stanton, N.A., Salmon, P.M., Walker, G.H., Baber, C. and Jenkins, D. 2005. *Human Factor Methods: A Practical Guide for Engineering and Design*. Aldershot, UK: Ashgate.

Stanton, N.A., Salmon, P.M., Walker, G.H. and Jenkins, D. 2009a. Genotype and phenotype schemata as models of situation awareness in dynamic command and control teams. *International Journal of Industrial Ergonomics*, 39, 480–489.

Stanton, N.A., Salmon, P.M., Walker, G.H. and Jenkins, D. 2009b. Genotype and phenotype schemata and their role in distributed situation awareness in collaborative systems. *Theoretical Issues in Ergonomics Science*, 10(1), 43–68.

Stanton, N.A., Salmon, P.M., Walker, G.H. and Jenkins, D.P. 2010b. Is situation awareness all in the mind? *Theoretical Issues in Ergonomics Science*, 11(1–2), 29–40.

Stanton, N.A. and Stammers, R.B. 2008. Bartlett and the future of ergonomics. *Ergonomics*, 51(1), 1–13.

Stanton, N.A., Stewart, R., Harris, D., Houghton, R.J., Baber, C., McMaster, R., Salmon, P. et al. 2006. Distributed situation awareness in dynamic systems: Theoretical development and application of an ergonomics methodology. *Ergonomics*, 49(12), 1288–1311.

Stanton, N.A. and Walker, G.H. 2011. Exploring the psychological factors involved in the Ladbroke Grove rail accident. *Accident Analysis and Prevention*, 43, 1117–1127.

Stanton, N.A. and Young, M.S. 1998. Is utility in the mind of the beholder? A study of ergonomics methods. *Applied Ergonomics*, 29(1), 41–54.

Stanton, N.A. and Young, M.S. 1999. What price ergonomics? *Nature*, 399, 197–198.

Stanton, N.A. and Young, M.S. 2000. A proposed psychological model of driving automation. *Theoretical Issues in Ergonomics Science*, 1(4), 315–331.

Stewart, R., Stanton, N.A., Harris, D., Baber, C., Salmon, P.M., Mock, M., Tatlock, K., Wells, L. and Kay, A. 2008. Distributed situation awareness in an airborne warning and control system: Application of novel ergonomics methodology. *Cognition, Technology and Work*, 10(3), 221–229.

Strauss, A. and Corbin, J. 1990. *Basics of Qualitative Research: Grounded Theory Procedures and Techniques*. London, UK: Sage.

Taynor, J., Crandall, B. and Wiggins, S. 1987. The reliability of the critical decision method. KATR-863(B)-87-07F. Prepared under Contract MDA903-86-C-0170 for the U.S. Army Research Institute Field Unit. Leavenworth, WA: Klein Associates, Inc.

Turner, B.A. 1978. *Man-Made Disasters*. London, UK: Wykeham Science Press.

Walker, G.H. 2005. Verbal protocol analysis. In N.A. Stanton, A. Hedge, K. Brookhuis, E. Salas and H. Hendrick (eds.), *Handbook of Human Factors and Ergonomics Methods*, pp. 30.1–30.8. Boca Raton, FL: CRC Press.

Walker, G.H., Gibson, H., Stanton, N.A., Baber, C., Salmon, P. and Green, D. 2006. Event Analysis of Systemic Teamwork (EAST): A novel integration of ergonomics methods to analyse C4i activity. *Ergonomics*, 49(12–13), 1345–1369.

Walker, G.H., Stanton, N.A., Baber, C., Wells, L., Gibson, H., Salmon, P.M. and Jenkins, D. 2010. From ethnography to the EAST method: A tractable approach for representing distributed cognition in Air Traffic Control. *Ergonomics*, 53 (2), 184–197.

Walker, G.H., Stanton, N.A. and Salmon, P.M. 2011. Cognitive compatibility of motorcyclists and car drivers. *Accident Analysis and Prevention*, 43(3), 878–888.

Walker, G.H., Stanton, N.A., Salmon, P.M. and Jenkins, D.P. 2009a. An evolutionary approach to network enabled capability. *International Journal of Industrial Ergonomics*, 39, 303–312.

Walker, G.H., Stanton, N.A., Stewart, R., Jenkins, D., Wells, L., Salmon, P. and Baber, C. 2009b. Using an integrated methods approach to analyse the emergent properties of military command and control. *Applied Ergonomics*, 40, 636–647.

Wason, P.C. 1960. On the failure to eliminate hypotheses in a conceptual task. *Quarterly Journal of Experimental Psychology*, 12, 129–140.

Wiegmann, D. and Boquet, A. 2005. Human error and general aviation accidents: A comprehensive, fine-grained analysis using HFACS. Report Number DOT/FAA/AM-05/24. Washington, DC: FAA.

Wiegmannm, D.A. and von Thaden, T.L. 2003. Using schematic aids to improve recall in incident reporting: The Critical Event Reporting Tool (CERT). *The International Journal of Aviation Psychology*, 13(2), 153–171.

Wilson, J.R. and Rutherford, A. 1989. Mental models: Theory and application in human factors. *Human Factors*, 31, 617–634.

Wilson, K.A., Salas, E., Priest, H.A. and Andrews, D. 2007. Errors in the heat of battle: Taking a closer look at shared cognition breakdowns through teamwork. *Human Factors: The Journal of the Human Factors and Ergonomics Society*, 49, 243–256.

Wong, W., Sallis, P. and O'Hare, D. 1997. Eliciting information portrayal requirements: Experiences with the critical decision method. In H. Thimbleby, B. O'Conaill and P. Thomas (eds.), *People and Computers XII, HCI '97 Conference of the British Computer Society Special Interest Group on Human–Computer Interaction*, pp. 397–415. Bristol, UK: Springer.

Woods, D.D., Dekker, S.W.A., Cook, R., Johannesen, L. and Sarter, N. 2010. *Behind Human Error.* Aldershot, UK: Ashgate.

Young, J.E., Klosko, J.S. and Weishaar, M.E. 2003. *Schema Therapy: A Practitioner's Guide.* New York: The Guildford Press.

Young, K.L., Lenne, M.G., Beanland, V., Salmon, P. and Stanton, N.A. 2015. Where do novice and experienced drivers direct their attention on approach to urban rail level crossings. *Accident Analysis and Prevention*, 77, 1–11.

Yu, X., Lau, E., Vicente, K.J. and Carter, M.W. 2002. Toward theory driven, quantitative performance measurement in ergonomics science: The abstraction hierarchy as a framework for data analysis. *Theoretical Issues in Ergonomics Science*, 3(2), 124–142.

Zangerwell, O.L. 1970. Sir Frederic Bartlett (1886–1969). *Quarterly Journal of Experimental Psychology*, 22, 77–81.

Zangwill, O.L. 1970. Obituary notice: Sir Frederic Bartlett (1886–1969). *Quarterly Journal of Experimental Psychology,* 22(2), 77–81.

Zhang, T., Kaber, D. and Hsiang, S. 2010. Characterisation of mental models in a virtual reality-based multitasking scenario using measures of situational awareness. *Theoretical Issues in Ergonomics Science*, 11(1–2), 99–118.

Author Index

J

K

L

M

N

O

P

R

S

T

W

Y

Z

Subject Index

D

E

K

L

M

N

O

P

R

T

U

V

W